T0349831

POLARIZATION PHENOMENA IN PHYSICS

Application To Nuclear Reactions

POLARIZATION PHENOMENA IN PHYSICS

Application To Nuclear Reactions

Makoto Tanifuji

Hosei University, Japan

World Scientific

NEW JERSEY · LONDON · SINGAPORE · BEIJING · SHANGHAI · HONG KONG · TAIPEI · CHENNAI · TOKYO

Published by

World Scientific Publishing Co. Pte. Ltd.

5 Toh Tuck Link, Singapore 596224

USA office: 27 Warren Street, Suite 401-402, Hackensack, NJ 07601

UK office: 57 Shelton Street, Covent Garden, London WC2H 9HE

British Library Cataloguing-in-Publication Data
A catalogue record for this book is available from the British Library.

POLARIZATION PHENOMENA IN PHYSICS
Applications to Nuclear Reactions

ISBN 978-981-3230-88-0

For any available supplementary material, please visit
http://www.worldscientific.com/worldscibooks/10.1142/10731#t=suppl

Typeset by Stallion Press
Email: enquiries@stallionpress.com

Printed in Singapore

Contents

Preface

Polarization phenomena in nuclear reactions are an indispensable source of information on spin-dependent interactions between the related particles such as projectiles and their targets, since the polarization means deviations from the uniform distribution of the spin direction of the particle and accordingly the polarization of the particles is induced by their spin-dependent interactions. This book provides the theoretical foundation of the polarization and as applications describes characteristics of the spin-dependent interactions obtained from polarization observables in some nuclear scattering.

Investigations of the interactions which include spin-dependent parts have made important contributions in clarifying mechanisms of nuclear reactions and also have opened new aspects on nuclear structure studies. For example, in the nuclear shell model the spin-orbit component of one-body potential for a nucleon has played the decisive role in the derivation of the well-known magic number [1]. Later, this spin-orbit interaction has been justified by the measurement of the polarization of protons scattered in elastic scattering by nuclei.

The polarization of the particle emitted in scattering is a useful observable as was stated above. However, its measurement is not always easy to perform, because the experiment is carried essentially by a double scattering, since to measure the polarization of the emitted particle, one has to treat one more scattering in addition to the original scattering. To avoid such difficulties, interests have been focused on an alternative method by the use of polarized beams or polarized targets, since in elastic scattering the polarization of the emitted particle for the unpolarized beam is equivalent to the analyzing power which describes the deviation of the cross section due to the polarization of the incident beam. There we treat only the single scattering. Then the analyzing power of reactions is now one of the important subjects in the polarization phenomena of nuclear reactions.

In scattering of composite particles by nuclei, virtual excitations of the projectile produce additional interactions between the projectile and the target nucleus. In the case of deuteron scattering, the spin-dependent part of this interaction is a T_L-type tensor interaction, while in the case of ^{7}Li scattering the spin-dependent part is the spin-orbit interaction. These features of the induced interaction reflect the characteristics of the internal structure of the given projectile and can be identified by the measurements of the analyzing powers of the polarized beam of the projectile.

One of important applications of the polarization phenomena is the determination of the spin parity of resonances. A resonance is specified by its spin and parity. Then in resonance reactions, the reaction amplitudes are restricted by the spin parity of the resonance. That is, the possible amplitude is the one in which the total angular momentum and parity are respectively the same as the resonance spin and parity. Such restrictions on the reaction amplitude induce polarizations or analyzing powers of the related particles. Conversely, the measurements of polarization observables can be used for the determination of the resonance spin.

Finally, we will emphasize that the study of the polarization phenomena encounters with serious difficulties in few nucleon systems. In proton-deuteron elastic scattering, the analyzing powers calculated by nucleon-nucleon interactions such as the AV18 one do not reproduce the measured analyzing powers satisfactorily. For example, the calculated vector analyzing power of the proton, A_y and that of the deuteron, iT_{11} at low energies reproduce the angular distributions of the measured analyzing powers but the theoretical magnitudes are much smaller compared with the measured. The calculation has been performed by several methods, for example by the Faddeev one or by variational ones, but their results have essentially the same nature and have not succeeded in reproducing the experimental data. This means that our nuclear interactions are not sufficient for understanding the polarization phenomena in three nucleon systems. Similar difficulties are also found in four nucleon systems. In scattering of protons by ^{3}He nuclei, the calculated Ay of the proton is almost half of the measured one at $Ep = 4{:}05$ MeV. These difficulties are named as "A_y puzzle" and can not be solved by introducing recently developed three nucleon forces due to exchanges of two pions between three nucleons. Now we have to look for a new idea of the interaction between nucleons.

More details on these subjects will be given in the following chapters of this book.

Chapter 1

Polarization, Alignment and Orientation

In this chapter, we will present the concept of polarizations in a naive sense, and describe the alignment of spin directions as an extension. Further, a quantum-mechanical description of the polarization is given for a spin $\frac{1}{2}$ particle by using its wave function.

1.1 Polarization, Alignment and Orientation

Let us consider an assembly of particles of spin s, which is quantized along the z axis. For simplicity, in the following, considerations are restricted tentatively to the case of $s = \frac{1}{2}$ and we denote the number of the particle which has $s_z = \pm\frac{1}{2}$ by $n_{\pm\frac{1}{2}}$. Then the total number of the particles of the assembly, n, is given by

$$n = n_{\frac{1}{2}} + n_{-\frac{1}{2}} \tag{1.1}$$

and the expectation value of the z component of the total spin of the system, $\langle s_z \rangle_{\text{assembly}}$ will be given by

$$\langle s_z \rangle_{\text{assembly}} = \frac{1}{2} n_{\frac{1}{2}} - \frac{1}{2} n_{-\frac{1}{2}}. \tag{1.2}$$

When the terms "spin-up" and "spin-down" are introduced for $s_z = \frac{1}{2}$ and $-\frac{1}{2}$, Eq. (1.2) means that the contribution of the spin-up particles, $\frac{1}{2} n_{\frac{1}{2}}$ is reduced by the contribution of the spin-down particles $-\frac{1}{2} n_{-\frac{1}{2}}$. In the sense of the average, the expectation value

1

of the z component of the spin of one particle $\langle s_z \rangle_{\text{particle}}$ is given by

$$\langle s_z \rangle_{\text{particle}} = \frac{1}{n} \langle s_z \rangle_{\text{assembly}} = \frac{1}{n} \left(\frac{1}{2} n_{\frac{1}{2}} - \frac{1}{2} n_{-\frac{1}{2}} \right) = \frac{1}{2} (q_{\frac{1}{2}} - q_{-\frac{1}{2}})$$
(1.3)

with

$$q_{\pm\frac{1}{2}} = \frac{1}{n} n_{\pm\frac{1}{2}},$$
(1.4)

where $q_{\frac{1}{2}}$ describes the probability of occupation by the spin-up state and $q_{-\frac{1}{2}}$ the one by the spin-down state. Inserting Eq. (1.4) into (1.1), we get

$$q_{\frac{1}{2}} + q_{-\frac{1}{2}} = 1,$$
(1.5)

which gives the normalization of q's.

We will define the polarization of the particle, p in terms of q's as

$$p = q_{\frac{1}{2}} - q_{-\frac{1}{2}},$$
(1.6)

which describes the effective spin-up probability for one particle.

For the general case of the spin s, the definition of the polarization, (1.6) is extended to become

$$p = \frac{1}{s} \sum_{\nu=-s}^{s} \nu q_\nu,$$
(1.7)

where ν is the z component of the spin s and q_ν describes the probability that s_z takes a particular value ν. The normalization of q_ν is given by

$$\sum_{\nu=-s}^{s} q_\nu = 1.$$
(1.8)

By fixing $s = \frac{1}{2}$, Eqs. (1.7) and (1.8) give Eqs. (1.6) and (1.5) as a special case.

The polarization is a measure of the distribution of the spin direction. In particular, when the distribution is uniform on ν, the polarization should vanish. This can be seen by setting $q_\nu = \bar{q}$, where $\bar{q}$ is a constant independent of ν. In this case, from Eq. (1.7),

$$p = \frac{\bar{q}}{s} \sum_{\nu=-s}^{s} \nu, \qquad (1.9)$$

which vanishes due to $\sum_{\nu=-s}^{s} \nu = 0$.

Next, we will consider a case different from the above, where q_ν depends on $|\nu|$ but $q_{-\nu} = q_\nu$. This happens often in experiments for $s \geq 1$. In this case, Eq. (1.7) gives $p = 0$. However, the distribution of the spin direction is not always uniform on ν. To describe such distributions, we will define the alignment A_l for $s \geq 1$ [2] by

$$A_l = \frac{1}{s^2} \sum_{\nu=-s}^{s} \left\{ \nu^2 - \frac{s(s+1)}{3} \right\} q_\nu, \qquad (1.10)$$

where the second term in the parentheses is chosen so that A_l vanishes when q_ν does not depend on ν. From Eq. (1.7), the polarization can be regarded as the dipole moment of the distribution of q_ν due to the weight ν in $\sum_\nu$. In this sense, the alignment (1.10) will be the quadrupole moment due to the weight ν^2. In other words, in the spin space, the polarization has the nature of a vector, while the alignment has that of a second-rank tensor. At present, we adopt a change of the terminology. Hereafter, we call p in Eq. (1.7) the vector polarization. Also, the word "polarization" is used to refer to the distribution of the spin direction q_ν, in general. In addition, the term "orientation", which is used often in experiments [2], refers to both the vector polarization and the alignment.

1.2 Quantum Mechanical Treatment of Vector Polarization

In this section, we will describe the vector polarization of the spin $\frac{1}{2}$ particle quantum-mechanically by using wave functions of the particle. Since both the spin-up and the spin-down configurations are mixed in the state we are considering, the wave function of the particle, ψ, is given by

$$\psi = a x_{\frac{1}{2}} + b x_{-\frac{1}{2}}, \qquad (1.11)$$

where $x_{\frac{1}{2}}$ is the spin function of the particle for the spin-up state and $x_{-\frac{1}{2}}$ is for the spin-down state. Since the wave function ψ is normalized to 1, we get

$$|a|^2 + |b|^2 = 1. \tag{1.12}$$

The spin of the particle s is given by $s = \frac{1}{2}\sigma$, where σ is the Pauli spin operator,

$$\sigma_x = \begin{pmatrix} 0 & 1 \\ 1 & 0 \end{pmatrix}, \quad \sigma_y = \begin{pmatrix} 0 & -i \\ i & 0 \end{pmatrix}, \quad \sigma_z = \begin{pmatrix} 1 & 0 \\ 0 & -1 \end{pmatrix}. \tag{1.13}$$

The spin functions $x_{\pm\frac{1}{2}}$ are the eigen functions of s_z. Then the expectation value of s_z in the ψ state is given by

$$\langle\psi|s_z|\psi\rangle = \frac{1}{2}(|a|^2 - |b|^2). \tag{1.14}$$

Comparing Eq. (1.14) with Eq. (1.3), we get

$$|a|^2 = q_{\frac{1}{2}} \quad \text{and} \quad |b|^2 = q_{-\frac{1}{2}}. \tag{1.15}$$

Then Eq. (1.6) is now expressed, by using p_z instead of p, as

$$p_z = 2\langle\psi|s_z|\psi\rangle. \tag{1.16}$$

This allows us to extend the definition of the polarization to the vector quantity

$$p = \langle\psi|\sigma|\psi\rangle, \tag{1.17}$$

where p_x, p_y, p_z are described by a and b as

$$p_x = 2\operatorname{Re}(a^*b), \tag{1.18}$$

$$p_y = 2\operatorname{Im}(a^*b) \tag{1.19}$$

$$\text{and} \quad p_z = |a|^2 - |b|^2. \tag{1.20}$$

These will help us for the determination of a and b. In the following chapters, we will discuss the polarization for the general spin by using a spin density matrix.

Chapter 2

Spin Observables by Density Matrix

As was shown in the previous chapter, in order to describe the distribution of the spin direction of a particle with spin s, spin space vectors are sufficient for $s = \frac{1}{2}$, but for $s = 1$, second-rank tensors are needed in addition to the vectors. In later sections, we will treat $s = \frac{3}{2}$ particles like ^{7}Li beams, for which higher-rank tensor quantities of the spin space will be required to describe the spin direction distribution. For the general case of arbitrary spins, it will be convenient to utilize the spin density matrix, the elements of which describe the distribution of the spin direction.

2.1 Density Matrix in Spin Space

Let us start from Eq. (1.11), the extension of the wave function of a particle with $s = \frac{1}{2}$, which describes the mixed state of the spin z component, to the case of arbitrary spin s. As a general case, we consider a wave function which describes an eigen state of observables except for the spin, which is denoted by ψ_α with the quantum number α. Equation (1.11) becomes [2]

$$\psi_\alpha = \sum_\nu C_{\alpha\nu}\phi_\alpha x_\nu, \qquad (2.1)$$

where ϕ_α is the eigen function of the observables except for the spin and x_ν is the spin wave function with the z component ν.

Next, we will consider the expectation value of a spin observable A in this state:

$$\langle A \rangle_\alpha \equiv \langle \psi_\alpha | A | \psi_\alpha \rangle = \sum_{\nu\nu'} C_{\alpha\nu'}^* C_{\alpha\nu} \langle \nu' | A | \nu \rangle. \tag{2.2}$$

Since $|C_{\alpha\nu}|^2$ specifically describes the density of the ν-state, we will extend the concept of the density to the general case by including the mixed state [2]. For that purpose, we will define the spin density matrix ρ by the matrix elements

$$\langle \nu | \rho | \nu' \rangle \equiv C_{\alpha\nu} C_{\alpha\nu'}^*, \tag{2.3}$$

where the diagonal element $\langle \nu | \rho | \nu \rangle$ gives the density of the ν-state. The expectation value of A is now written as

$$\langle A \rangle_\alpha = \sum_{\nu\nu'} \langle \nu | \rho | \nu' \rangle \langle \nu' | A | \nu \rangle = Tr(\rho A), \tag{2.4}$$

where the operator Tr refers to the sum over the diagonal matrix elements of ρA.

Next, we will examine properties of ρ in the spin space. In general, ρ will be expanded in the spherical tensors of the spin space, τ_κ^k, where k is the rank and κ is the z component

$$\rho = \frac{1}{2s+1} \sum_{k\kappa} (-)^\kappa t_{k\kappa} \tau_{-\kappa}^k. \tag{2.5}$$

Since ρ is a scalar quantity, the expansion coefficient $t_{k\kappa}$ is also a tensor designated by the rank k and the z component κ. The matrix element of τ_κ^k is given by the Wigner–Eckart theorem [3]. Introducing the spin s explicitly in the definition of the state,

$$\langle s\nu | \tau_\kappa^k | s\nu' \rangle = \frac{1}{\sqrt{2s+1}} (sk\nu'\kappa \mid s\nu)(s\|\tau^k\|s). \tag{2.6}$$

Here $\langle sk\nu'\kappa \mid s\nu \rangle$ is the Clebsch–Gordan (hereafter abbreviated by CG) coefficient [3] for the addition of two vectors s and k when the resultant vector is s, and ν', κ and ν are their z components, respectively. $(s\|\tau^k\|s)$ is the physical part of the matrix element. At present, the magnitude of τ_κ^k is normalized to give

$$(s\|\tau^k\|s) = \sqrt{(2s+1)(2k+1)}.$$ (2.7)

Hence

$$\langle s\nu|\rho|s\nu'\rangle = \frac{1}{2s+1}\sum_{k\kappa}(-)^\kappa\sqrt{2k+1}(sk\nu' - \kappa \mid s\nu)t_{k\kappa}.$$ (2.8)

The nature of the CG coefficient $(sk\nu' - \kappa \mid s\nu)$ limits k to

$$0 \le k \le 2s$$ (2.9)

and leads to

$$\tau_\kappa^k = (-)^\kappa\tau_{-\kappa}^k{}^\dagger.$$ (2.10)

Equation (2.9) indicates that the density matrix consists of $k = 0$ and $k = 1$ terms for $s = \frac{1}{2}$, and $k = 0$, $k = 1$ and $k = 2$ terms for $s = 1$. That is, the polarization is described by the vector for $s = \frac{1}{2}$ and by the vector and the second-rank tensor for $s = 1$.

Next, we will calculate the expectation value of the tensor operator in the spin space, τ_κ^k, for a given ρ by setting $A = \tau_\kappa^k$ in Eq. (2.4). Using Eq. (2.6) and the orthogonality condition of the CG coefficients [3], we get

$$\langle\tau_\kappa^k\rangle = Tr(\rho\tau_\kappa^k) = t_{k\kappa}.$$ (2.11)

This means that $t_{k\kappa}$, introduced as the expansion coefficient of ρ, gives the expectation value of τ_κ^k in the given density ρ. That is, $t_{k\kappa}$ describes the magnitude of the τ_κ^k component in the distribution of the spin direction.

Further, we will investigate a special case where ρ has only diagonal elements, which we call the pure state, and denote the density by $\bar{\rho}$. This $\bar{\rho}$ will be obtained from ρ by a rotation of the coordinate axes. An example of such rotations will be given in Sec. 3.4. At present, Eq. (2.3) is reduced to

$$\langle \nu | \bar{\rho} | \nu' \rangle = |C_{\alpha\nu}|^2 \delta\nu'\nu, \tag{2.12}$$

where $|C_{\alpha\nu}|^2$ expresses the probability of finding the system in the $|s\nu\rangle$ state and is equivalent to q_ν introduced in Eq. (1.7). Then we write

$$\langle \nu | \bar{\rho} | \nu' \rangle = q_\nu \delta\nu'\nu. \tag{2.13}$$

Further, $t_{k\kappa}$ in the pure state, which we denote as $\bar{t}_{k\kappa}$, is obtained from Eqs. (2.11) and (2.6) to be

$$\bar{t}_{k\kappa} \begin{cases} = 0 & \text{for } \kappa \neq 0 \\ = Tr(\bar{\rho}\tau_0^k) = \sqrt{2k+1} \sum_\nu (sk\nu0 \mid s\nu)q_\nu & \text{for } \kappa = 0, \end{cases} \tag{2.14}$$

which gives

(i)   for $s = \frac{1}{2}$

$$\bar{t}_{10} = q_{\frac{1}{2}} - q_{-\frac{1}{2}}, \tag{2.15}$$

(ii)   for $s = 1$

$$\bar{t}_{10} = \sqrt{\frac{3}{2}}(q_1 - q_{-1}). \tag{2.16}$$

These results are equivalent to Eq. (1.7) since $p_z = \sqrt{\frac{2}{3}}\tau_0^1$ for $s = 1$, as will be discussed in the next section. Finally, because of $\tau_0^0 = 1$, Eq. (2.11) gives

$$t_{00} = Tr(\rho). \tag{2.17}$$

This suggests that $t_{k\kappa}$ will be normalized by using

$$t_{k\kappa} = \frac{1}{Tr(\rho)} Tr(\rho\tau_\kappa^k), \tag{2.18}$$

when ρ is not normalized, for example, due to reactions. This will be discussed in more detail in the next chapter.

2.2 Relationships between Representations, Spherical and Cartesian

In the previous chapter, we adopted the Cartesian coordinate system to describe the physical quantities, while in this chapter we have adopted the spherical coordinate system. The relation between both systems has already been examined [4] and we will summarize the results of ref. [5].

The operators in the Cartesian coordinate system are described by P_i with $i = x, y, z$ and P_{ij} with $ij = xx, yy, zz, xy, yz, zx$, etc. The operators in the spherical coordinate system are described by τ_κ^k. The Cartesian coordinates are defined by referring to the momenta of the incident and outgoing particles of a reaction, $\boldsymbol{k}_i$ and $\boldsymbol{k}_f$, where $\boldsymbol{k}_i \parallel z$-axis and $\boldsymbol{k}_i \times \boldsymbol{k}_f \parallel y$-axis, which is called the Madison Convention. The relations between the two systems are given as follows.

(i) In the case of $s = \frac{1}{2}$,

$$P_x = \frac{1}{\sqrt{2}}(-\tau_1^1 + \tau_{-1}^1), \tag{2.19}$$

$$P_y = \frac{i}{\sqrt{2}}(\tau_1^1 + \tau_{-1}^1), \tag{2.20}$$

$$P_z = \tau_0^1 \tag{2.21}$$

and solving Eqs. (2.19) and (2.20) we get

$$\tau_{\pm 1}^1 = \mp \frac{1}{\sqrt{2}}(P_x \pm iP_y). \tag{2.22}$$

(ii) In the case of $s = 1$,

for vector operators

$$P_x = \frac{1}{\sqrt{3}}(-\tau_1^1 + \tau_{-1}^1), \tag{2.23}$$

$$P_y = \frac{i}{\sqrt{3}}(\tau_1^1 + \tau_{-1}^1), \tag{2.24}$$

$$P_z = \sqrt{\frac{2}{3}}\tau_0^1 \tag{2.25}$$

and solving Eqs. (2.23), (2.24) and (2.25) we get

$$\tau^1_{\pm 1} = \mp \frac{\sqrt{3}}{2}(P_x \pm iP_y), \tag{2.26}$$

$$\tau^1_0 = \sqrt{\frac{3}{2}}P_z; \tag{2.27}$$

for tensor operators

$$P_{xx} = \frac{\sqrt{3}}{2}(\tau^2_2 + \tau^2_{-2}) - \frac{1}{\sqrt{2}}\tau^2_0, \tag{2.28}$$

$$P_{yy} = -\frac{\sqrt{3}}{2}(\tau^2_2 + \tau^2_{-2}) - \frac{1}{\sqrt{2}}\tau^2_0, \tag{2.29}$$

$$P_{zz} = \sqrt{2}\tau^2_0, \tag{2.30}$$

$$P_{xy} = -\frac{\sqrt{3}}{2}i(\tau^2_2 - \tau^2_{-2}), \tag{2.31}$$

$$P_{zx} = -\frac{\sqrt{3}}{2}(\tau^2_1 - \tau^2_{-1}), \tag{2.32}$$

$$P_{yz} = \frac{\sqrt{3}}{2}i(\tau^2_1 + \tau^2_{-1}), \tag{2.33}$$

where P_{ij} satisfy the relation

$$P_{xx} + P_{yy} + P_{zz} = 0, \quad P_{yx} = P_{xy}, \quad P_{xz} = P_{zx}, \quad P_{zy} = P_{yz}, \tag{2.34}$$

and solving Eqs. (2.28)–(2.33) we get

$$\tau^2_0 = \frac{1}{\sqrt{2}}P_{zz}, \tag{2.35}$$

$$\tau^2_{\pm 1} = \mp \frac{1}{\sqrt{3}}(P_{xz} \pm iP_{yz}), \tag{2.36}$$

$$\tau^2_{\pm 2} = \frac{1}{\sqrt{3}}\left\{\frac{1}{2}(P_{xx} - P_{yy}) \pm iP_{xy}\right\}. \tag{2.37}$$

(iii) The operators discussed above can be represented by the spin operator *s*.

For the $s = \frac{1}{2}$ case,

$$\boldsymbol{P} = \boldsymbol{\sigma} = 2\boldsymbol{s}. \tag{2.38}$$

For the $s = 1$ case,

$$P = s, \tag{2.39}$$

$$P_{xx} = 3s_x^2 - 2, \tag{2.40}$$

$$P_{yy} = 3s_y^2 - 2, \tag{2.41}$$

$$P_{zz} = 3s_z^2 - 2, \tag{2.42}$$

$$P_{xy} = \frac{3}{2}(s_x s_y + s_y s_x), \tag{2.43}$$

$$P_{xz} = \frac{3}{2}(s_x s_z + s_z s_x), \tag{2.44}$$

$$P_{yz} = \frac{3}{2}(s_y s_z + s_z s_y), \tag{2.45}$$

where $\boldsymbol{s}$ is given by the matrix designating the row and column by the eigenstate of s_z,

$$s_x = \frac{1}{\sqrt{2}} \begin{pmatrix} 0 & 1 & 0 \\ 1 & 0 & 1 \\ 0 & 1 & 0 \end{pmatrix}, \quad s_y = \frac{1}{\sqrt{2}} \begin{pmatrix} 0 & -i & 0 \\ i & 0 & -i \\ 0 & i & 0 \end{pmatrix},$$

$$s_z = \frac{1}{\sqrt{2}} \begin{pmatrix} 1 & 0 & 0 \\ 0 & 0 & 0 \\ 0 & 0 & -1 \end{pmatrix}. \tag{2.46}$$

For the $s = \frac{3}{2}$ case, τ_x^1 and τ_x^2 are given in terms of s_x, s_y and s_z as

$$\tau_{\pm 1}^1 = \mp \sqrt{\frac{2}{5}}(s_x \pm i s_y), \tag{2.47}$$

$$\tau_0^1 = \frac{2}{\sqrt{5}} s_z, \tag{2.48}$$

$$\tau_{\pm 2}^2 = \frac{1}{\sqrt{6}}\{(s_x^2 - s_y^2) \pm i(s_x s_y + s_y s_x)\}, \tag{2.49}$$

$$\tau^2_{\pm 1} = \mp \frac{1}{\sqrt{6}} \{(s_x s_z + s_z s_x) \pm i(s_y s_z + s_z s_y)\}, \qquad (2.50)$$

$$\tau^2_0 = s^2_z - \frac{5}{4}, \qquad (2.51)$$

where s is given by [6]

$$s_x = \frac{1}{2} \begin{pmatrix} 0 & \sqrt{3} & 0 & 0 \\ \sqrt{3} & 0 & 2 & 0 \\ 0 & 2 & 0 & \sqrt{3} \\ 0 & 0 & \sqrt{3} & 0 \end{pmatrix},$$

$$s_y = \frac{1}{2} \begin{pmatrix} 0 & -\sqrt{3}i & 0 & 0 \\ \sqrt{3}i & 0 & -2i & 0 \\ 0 & 2i & 0 & -\sqrt{3}i \\ 0 & 0 & \sqrt{3}i & 0 \end{pmatrix},$$

$$s_z = \frac{1}{2} \begin{pmatrix} 3 & 0 & 0 & 0 \\ 0 & 1 & 0 & 0 \\ 0 & 0 & -1 & 0 \\ 0 & 0 & 0 & -3 \end{pmatrix}. \qquad (2.52)$$

Chapter 3

Spin Observables
in Nuclear Reactions

3.1 General Formulae of Cross Section and Polarization in Nuclear Reactions

Since polarizations are induced by spin-dependent interactions, polarization phenomena in nuclear reactions are a very important tool to investigate spin-dependent interactions between related particles. In actual cases, to obtain information of such interactions, one will observe effects on the cross section of reactions due to polarizations of the incident beams and/or measure polarizations of emitted particles induced by the reaction. In the following, we will present fundamental formulae to treat such applications.

For convenience in the treatment of the polarization observables, we will introduce the spin density matrices, ρ_i and ρ_f, for the initial state and the final state, respectively. They describe the distributions of the spin direction of the related particles in the respective state. In general, quantities in the initial state are transformed by the transition matrix of reaction, M, to those in the final state [7]. Then ρ_f is obtained from ρ_i by M as

$$\rho_f = M \rho_i M^\dagger. \tag{3.1}$$

Let us consider a reaction $a + A \rightarrow b + B$, where a and b are the incident and emitted particles and A and B are the target and residual nuclei. In the spin space, the final state of the reaction will be designated by the spin z component of b and B, say ν_b and ν_B.

13

Then the diagonal matrix element of ρ_f, $\langle \nu_b \nu_B | \rho_f | \nu_b \nu_B \rangle$ describes the probability of finding the system in $|\nu_b \nu_B\rangle$ state as was discussed in the previous chapter. Choosing the constant factor properly, we will write the differential cross section of the reaction $\frac{d\sigma}{d\Omega}$ as

$$\frac{d\sigma}{d\Omega} = \frac{\mu_{aA}\mu_{bB}}{(2\pi)^2} \frac{k_f}{k_i} Tr(\rho_f), \qquad (3.2)$$

where μ_{aA} and μ_{bB} are the reduced masses, and $\mathbf{k}_i$ and $\mathbf{k}_f$ are the relative momenta, in the initial and final states, respectively. The polarization of the emitted particle, $t_{k_b \kappa_b}$, is given by Eq. (2.18). Considering that the present ρ_f is not normalized,

$$t_{k_b \kappa_b} = \frac{1}{Tr(\rho_f)} Tr(\rho_f \tau_{\kappa_b}^{k_b}). \qquad (3.3)$$

Applications to special cases will be given in the subsequent sections, where Eqs. (3.1)–(3.3) will be found to agree with the conventional formulae.

3.2 Cross Section and Polarization for Unpolarized Beam and Target

Let us consider the case where the beam and the target are not polarized. The normalized initial spin density ρ_i is given as

$$\rho_i = \frac{1}{(2s_a + 1)(2s_A + 1)}. \qquad (3.4)$$

Thus from Eq. (3.1)

$$\rho_f = \frac{1}{(2s_a + 1)(2s_A + 1)} MM^\dagger. \qquad (3.5)$$

Inserting Eqs. (3.4) and (3.5) into (3.2), we obtain the differential cross section for the unpolarized beam and target, $\frac{d\sigma}{d\Omega}\big|_0$ as

$$\frac{d\sigma}{d\Omega}\bigg|_0 = \frac{\mu_{aA}\mu_{bB}}{(2\pi)^2} \frac{k_f}{k_i} \frac{1}{(2s_a + 1)(2s_A + 1)}$$

$$\times \sum_{\nu_a \nu_A \nu_b \nu_B} |\langle \nu_p \nu_B; \mathbf{k}_f | M | \nu_a \nu_A; \mathbf{k}_i \rangle|^2, \qquad (3.6)$$

where ν's are the z components of the spins. Equation (3.6) agrees with the conventional formula, which adds validity to the development of Eqs. (3.1) and (3.2).

Next, we will calculate the polarization of the emitted particle b. From Eq. (3.3)

$$t_{k_b \kappa_b} = \frac{1}{N_R} \text{Tr}(MM^\dagger \tau_{\kappa_b}^{k_b}), \qquad (3.7)$$

where

$$N_R = \text{Tr}(MM^\dagger). \qquad (3.8)$$

In the following, we will derive the explicit formula of p_y for $s_b = \frac{1}{2}$. From Eq. (3.7) we get

$$t_{11} = -\frac{\sqrt{2}}{N_R} \sum_{\nu_a \nu_A \nu_B} \left\langle \nu_b = -\frac{1}{2}, \nu_B; \boldsymbol{k}_f | M | \nu_a \nu_A; \boldsymbol{k}_i \right\rangle$$

$$\times \left\langle \nu_b = \frac{1}{2}, \nu_B; \boldsymbol{k}_f | M | \nu_a \nu_A; \boldsymbol{k}_i \right\rangle^*, \qquad (3.9)$$

for which one can change the signs of ν_a, ν_A and ν_B since these variables run over all values in the operation of $\sum_{\nu_a \nu_A \nu_B}$. Then we get

$$t_{11} = -\frac{\sqrt{2}}{N_R} \sum_{\nu_a \nu_A \nu_B} \frac{1}{2} \left\{ \left\langle \nu_b = -\frac{1}{2}, \nu_B; \boldsymbol{k}_f | M | \nu_a \nu_A; \boldsymbol{k}_i \right\rangle \right.$$

$$\times \left\langle \nu_b = \frac{1}{2}, \nu_B; \boldsymbol{k}_f | M | \nu_a \nu_A; \boldsymbol{k}_i \right\rangle^*$$

$$+ \left\langle \nu_b = -\frac{1}{2}, -\nu_B; \boldsymbol{k}_f | M | -\nu_a -\nu_A; \boldsymbol{k}_i \right\rangle$$

$$\left. \times \left\langle \nu_b = \frac{1}{2}, -\nu_B; \boldsymbol{k}_f | M | -\nu_a -\nu_A; \boldsymbol{k}_i \right\rangle^* \right\}. \qquad (3.10)$$

Further, from [8],

$$\langle -\nu_b -\nu_B | M | -\nu_a -\nu_A \rangle = (-)^{\Delta\pi+\Delta P} \langle \nu_b \nu_B | M | \nu_a \nu_A \rangle, \qquad (3.11)$$

where ΔP is the parity change in this reaction and

$$\Delta\pi = (s_a + s_A + \nu_a + \nu_A) - (s_b + s_B + \nu_b + \nu_B). \qquad (3.12)$$

Using Eqs. (3.10) and (3.11), we obtain

$$it_{11} = \frac{\sqrt{2}}{N_R}\mathrm{Im}\left\{ \sum_{\nu_a\nu_A\nu_B} \left\langle \nu_b = \frac{1}{2}, \nu_B; \boldsymbol{k}_f|M|\nu_a\nu_A; \boldsymbol{k}_i \right\rangle^* \right.$$

$$\left. \times \left\langle \nu_b = -\frac{1}{2}, \nu_B; \boldsymbol{k}_f|M|\nu_a\nu_A; \boldsymbol{k}_i \right\rangle \right\}. \qquad (3.13)$$

Similarly we get

$$it_{1-1} = it_{11}. \qquad (3.14)$$

Equation (2.20) gives

$$p_y = \sqrt{2}it_{11}. \qquad (3.15)$$

When Eq. (3.13) is inserted, Eq. (3.15) gives the full expression for the vector polarization formula (1.19).

3.3 Cross Section and Analyzing Power for Polarized Beam or Target

In this section we will investigate the effects on the cross section due to polarization of the incident beam or the target nucleus. Further, the investigation will be focused on the case of the polarized beam, because, theoretically, the investigation of the polarized target is equivalent to that of the polarized beam when the direction of the incident momentum is reversed. For polarized beams on unpolarized targets, the density matrices are given as

$$\rho_a = \frac{1}{2s_a + 1}\sum_{k_a\kappa_a}(-)^{\kappa_a}t_{k_a\kappa_a}\tau^{k_a}_{-\kappa_a} \qquad (3.16)$$

and

$$\rho_A = \frac{1}{2s_A + 1}. \qquad (3.17)$$

The cross section is given by

$$\frac{d\sigma}{d\Omega} = \frac{\mu_{aA}\mu_{bB}}{(2\pi)^2} \frac{k_f}{k_i} \frac{1}{(2s_a+1)(2s_A+1)} Tr\left(M \sum_{k_a\kappa_a} (-)^{\kappa_a} t_{k_a-\kappa_a} \tau_{\kappa_a}^{k_a} M^\dagger \right),$$

which one can write in a compact form by defining the analyzing power $T_{k_a\kappa_a}$,

$$T_{k_a\kappa_a} = \frac{1}{N_R} Tr(M\tau_{\kappa_a}^{k_a} M^\dagger). \tag{3.18}$$

This describes the effect of the beam polarization. Using this, we get

$$\frac{d\sigma}{d\Omega} = \frac{d\sigma}{d\Omega}\bigg|_0 \left\{ 1 + \sum_{k_a\neq 0}\sum_{\kappa_a} (-)^{\kappa_a} t_{k_a-\kappa_a} T_{k_a\kappa_a} \right\}. \tag{3.19}$$

Then the analyzing power $T_{k_a\kappa_a}$ will be obtained by the measurement of the cross section for the given beam polarization $t_{k_a\kappa_a}$. The analyzing power in the Cartesian representation, e.g. the vector one A_i and the tensor one A_{ij}, are defined similarly to Eq. (3.18),

$$A_i = \frac{1}{N_R} Tr(MP_i M^\dagger) \tag{3.20}$$

and

$$A_{ij} = \frac{1}{N_R} Tr(MP_{ij} M^\dagger), \tag{3.21}$$

where i and j are x, y or z, and P_i and P_{ij} are given in Chapter 2.

Let us examine in more detail the analyzing power for the $a+A \rightarrow b + B$ reaction with $s_a = \frac{1}{2}$. In this case, due to Eq. (2.9), only the vector polarization is allowed for the incident beam. Using Eq. (3.20), the cross section is written in the Cartesian representation as

$$\frac{d\sigma}{d\Omega} = \frac{d\sigma}{d\Omega}\bigg|_0 \left\{ 1 + \boldsymbol{p} \cdot \boldsymbol{A} \right\}, \tag{3.22}$$

where $\boldsymbol{p}$ is the polarization vector and $\boldsymbol{A}$ is the analyzing power given by (3.20). We will adopt the Madison Convention for the coordinate system, $z \parallel \boldsymbol{k}_i$ and $y \parallel \boldsymbol{k}_i \times \boldsymbol{k}_y$, which is shown in Fig. 3.1. First, we

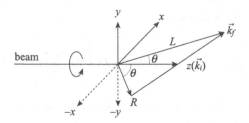

Fig. 3.1 Rotation of the coordinate axis around the z-axis by $180°$.

consider the case where the incident beam is fully polarized along the y-axis, i.e.

$$p_x = 0, \quad p_y = 1 \text{ and } p_z = 0.$$

For this case, Eq. (3.22) is reduced to

$$\left. \frac{d\sigma}{d\Omega} \right|_{\text{UP}} = \left. \frac{d\sigma}{d\Omega} \right|_0 (1 + A_y), \tag{3.23}$$

where the suffix UP means that the spin of the particle a is directed to the "up" side of the reaction plane. Secondly, we consider the case where the polarization of a is along the y-axis but is directed to the "down" side. In this case, Eq. (3.22) becomes

$$\left. \frac{d\sigma}{d\Omega} \right|_{\text{DOWN}} = \left. \frac{d\sigma}{d\Omega} \right|_0 (1 - A_y). \tag{3.24}$$

Further, in the second case, we will rotate the coordinate axes around the z-axis by $180°$. Through this rotation, the spin of the particle a is directed to the "up" side and the final particle b is emitted in the $(-x)z$ quarter of the reaction plane. Since the cross section itself is not influenced by the rotation, we will rewrite $\frac{d\sigma}{d\Omega}\big|_{\text{DOWN}}$ by the right-side cross section $\frac{d\sigma}{d\Omega}\big|_R$, which means the final particle b will be emitted in the right side with reference to the z axis. Accordingly, we rewrite $\frac{d\sigma}{d\Omega}\big|_{\text{UP}}$ using the left-side cross section $\frac{d\sigma}{d\Omega}\big|_L$. Usually, we use the left–right asymmetry A_{asym}, which is defined as

$$A_{\text{asym}} = \left\{ \left. \frac{d\sigma}{d\Omega} \right|_L - \left. \frac{d\sigma}{d\Omega} \right|_R \right\} \bigg/ \left\{ \left. \frac{d\sigma}{d\Omega} \right|_L + \left. \frac{d\sigma}{d\Omega} \right|_R \right\}. \tag{3.25}$$

Combined with Eqs. (3.23) and (3.24),

$$A_{\text{asym}} = A_y. \tag{3.26}$$

That is, the left–right asymmetry is equivalent to the vector analyzing power for $s_a = \frac{1}{2}$. Next, we will derive the explicit form of A_y for $s_a = \frac{1}{2}$. Since $P_y = \sigma_y$ for the spin $\frac{1}{2}$ particle, one can write

$$A_y = \frac{1}{N_R} Tr(MP_y M^\dagger)$$

$$= \frac{1}{N_R} Tr(M\sigma_y M^\dagger). \tag{3.27}$$

Using Eq. (3.27), we obtain

$$A_y = \frac{2}{N_R} \sum_{\nu_A} \sum_{\nu_b \nu_B} Im \left\{ \left\langle \nu_b \nu_B; \boldsymbol{k}_f | M | \nu_a = \frac{1}{2}\nu_A; \boldsymbol{k}_i \right\rangle \right.$$

$$\times \left. \left\langle \nu_b \nu_B; \boldsymbol{k}_f | M | \nu_a = -\frac{1}{2}\nu_A; \boldsymbol{k}_i \right\rangle^* \right\}. \tag{3.28}$$

Let us consider the reaction to be an elastic scattering process. In that case, the emitted particle b is the same as the incident particle a and we will show the vector analyzing power of the incident particle A_y to be equivalent to the polarization of the emitted particle p_y. The time reversal invariance of the scattering amplitude for the spin (z-component) $s(\nu)$ particle is given in ref. [8].

$$\langle s\nu'; \boldsymbol{k}_f | M | s\nu; \boldsymbol{k}_i \rangle = (-)^{\nu' - \nu} \langle s - \nu; -\boldsymbol{k}_i | M | s - \nu'; -\boldsymbol{k}_f \rangle. \tag{3.29}$$

Applying the above relation to Eq. (3.28), we get

$$A_y = \frac{2}{N_R} \sum_{\nu_B} \sum_{\nu_a \nu_A} Im \left\{ \left\langle \nu_b = \frac{1}{2}\nu_B; -\boldsymbol{k}_i | M | \nu_a \nu_A; -\boldsymbol{k}_f \right\rangle^* \right.$$

$$\times \left. \left\langle \nu_b = -\frac{1}{2}\nu_B; -\boldsymbol{k}_i | M | \nu_a \nu_A; -\boldsymbol{k}_f \right\rangle \right\}. \tag{3.30}$$

Considering that the amplitude is a function of the magnitude of the momentum and the scattering angle, one can understand the resultant A_y to be equivalent to p_y which is given by Eq. (3.15).

For $s_a = 1$ particles, Eqs. (2.26) and (2.27) for the vector polarization give

$$\sum_{\kappa_a} (-)^{\kappa_a} t_{1\kappa_a} T_{1-\kappa_a} = \frac{3}{2} (\boldsymbol{p} \cdot \boldsymbol{A}).$$

Then Eq. (3.19) becomes

$$\frac{d\sigma}{d\Omega} = \frac{d\sigma}{d\Omega}\bigg|_0 \left(1 + \frac{3}{2} \boldsymbol{p} \cdot \boldsymbol{A}\right). \tag{3.31}$$

For the full y-axis polarization of the beam,

$$A_{\text{asym}} = \frac{3}{2} A_y = \sqrt{3} i T_{11}. \tag{3.32}$$

3.4 Analyzing Power for Aligned Beam

In this section, we discuss the theoretical aspects of the analyzing power for aligned beams. The application to a realistic case will be given for ^{7}Li scattering in Chapter 7. For the convenience of experiments, the coordinate system EXP is introduced, where $y \parallel \boldsymbol{k}_i$ and $z \parallel \boldsymbol{k}_i \times \boldsymbol{k}_f$. The polarization of the beam is given in the pure state of the rank k discussed in Sec. 2.1, which is given as

$$\bar{t}_{k\kappa} \begin{cases} = 1 & \text{for } \kappa = 0 \\ = 0 & \text{for } \kappa \neq 0. \end{cases} \tag{3.33}$$

The cross section of reactions is given by using the corresponding analyzing power, which we denote by $^T T_{k0}$:

$$\frac{d\sigma}{d\Omega} = \frac{d\sigma}{d\Omega}\bigg|_0 (1 + \bar{t}_{k0}{}^T T_{k0}). \tag{3.34}$$

Using Eq. (3.33) we get

$$^T T_{k0} = \left\{\frac{d\sigma}{d\Omega} - \frac{d\sigma}{d\Omega}\bigg|_0\right\} \bigg/ \frac{d\sigma}{d\Omega}\bigg|_0. \tag{3.35}$$

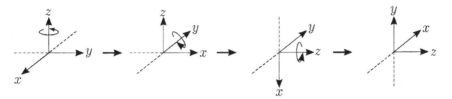

Fig. 3.2 Rotation of coordinate axes.

Next, we will transform the beam polarization by the rotation of the coordinate axes to the Madison Convention, $z \parallel \boldsymbol{k}_i$ and $y \parallel \boldsymbol{k}_i \times \boldsymbol{k}_f$. The transformation is performed by the use of the rotation matrix $D^k_{\kappa'\kappa}(\alpha, \beta, \gamma)$. The new polarization $t_{k\kappa}$ is given by

$$t_{k\kappa} = \sum_{\kappa'} \bar{t}_{k\kappa'} D^k_{\kappa'\kappa}(\alpha, \beta, \gamma), \tag{3.36}$$

where α, β and γ are the so-called Euler angles; that is, β is the rotation angle around the z-axis, α is the rotation angle around the new y-axis, and γ is the rotation angle around the new z-axis. The rotation will be carried out counter clockwise in the right-handed system. As is seen in Fig. 3.2, $\alpha = \beta = \gamma = \frac{\pi}{2}$ in the present case. Since

$$D^k_{o\kappa}\left(\frac{\pi}{2}, \theta, \frac{\pi}{2}\right) = i^\kappa \sqrt{\frac{4\pi}{2k+1}} Y_{k\kappa}(\theta, 0), \tag{3.37}$$

the polarized beams (3.33) in the EXP system are described by aligned beams in the Madison Convention system.

For $k = 1$,

$$t_{10} = 0, \quad t_{1\pm1} = -\frac{i}{\sqrt{2}}. \tag{3.38}$$

Calculating $\frac{d\sigma}{d\Omega}$ using Eq. (3.19) with the above $t_{1\kappa}$, we obtain

$$\frac{d\sigma}{d\Omega} = \frac{d\sigma}{d\Omega}\bigg|_0 (1 + \sqrt{2}iT_{11}).$$

Connecting this with Eq. (3.35), we get

$${}^T T_{10} = \sqrt{2}iT_{11}. \tag{3.39}$$

For $k = 2$,

$$t_{20} = -\frac{1}{2}, \quad t_{2\pm 1} = 0, \quad t_{2\pm 2} = -\sqrt{\frac{3}{8}}. \tag{3.40}$$

They give

$$^T T_{20} = -\frac{1}{2}(T_{20} + \sqrt{6}T_{22}). \tag{3.41}$$

For $k = 3$,

$$t_{3\pm 1} = i\sqrt{\frac{3}{16}}, \quad t_{3\pm 2} = 0, \quad t_{3\pm 3} = i\sqrt{\frac{5}{16}}. \tag{3.42}$$

They give

$$^T T_{30} = -\frac{1}{2}(\sqrt{3}iT_{31} + \sqrt{5}iT_{33}). \tag{3.43}$$

In Eqs. (3.38), (3.40) and (3.42), one will find that these beams satisfy the relation $t_{k-\kappa} = t_{k\kappa}$, which characterize the aligned beam as discussed in Sec. 1.1. Then, in the analysis of experimental data, for example in Chapter 7, we treat $^T T_{20}$ and $^T T_{30}$ as the analyzing power for the aligned beam. However, $^T T_{10}$ is described in terms of the conventional vector analyzing power iT_{11}.

3.5   Coefficients for Polarization Transfer, Depolarization and Spin Correlation

In this section, we will present the formulae which describe the effects of polarizations when two particles are polarized in a reaction $a+A \rightarrow b + B$ [5, 9]. For example, we consider the incident particle a to be polarized, as one of the two polarized particles.

When the polarization of the emitted particle b is measured, the feature of the reaction will be considered to be the transfer of the polarization from a to b. Then the characteristics of the reaction are described by the polarization transfer coefficient K. For example, the vector-to-vector polarization transfer coefficient is defined by

$$K_i^j = \frac{1}{N_R} Tr(MP_i(a)M^\dagger P_j(b)), \tag{3.44}$$

where $P_i(a)$ and $P_j(b)$ describe the vector polarization operator of the particles a and b, respectively, which are defined in Sec. 2.2. The suffices i and j denote the component of $\boldsymbol{P}$, i.e., x, y, or z. The coefficient for the transition from the ij tensor polarization to the k vector polarization is described by

$$K_{ij}^k = \frac{1}{N_R} Tr(MP_{ij}(a)M^\dagger P_k(b)) \tag{3.45}$$

and so on.

As a special case, the depolarization D_i^j is defined for scattering of the $s = \frac{1}{2}$ particle by the spinless target nucleus. The depolarization of the particle a $(s = \frac{1}{2})$ is defined as

$$D_i^j = \frac{1}{N_R} Tr(M\sigma_i(a)M^\dagger \sigma_j(a)), \tag{3.46}$$

where σ_i, σ_j are the i and j components of the Pauli spin $\boldsymbol{\sigma}$ of a, respectively, and i and j denote x, y, or z. Then, D_i^j becomes one type of the polarization transfer coefficient discussed above. In Chapter 9, we will analyze experimental data of D_i^j to find important contributions of the nucleon–nucleus spin-dependent interactions.

Experiments where both the beam and target are polarized have been attempted, in particular for the nucleon–nucleon scattering. The effect of the polarizations on the cross section is described by the spin correlation coefficient, $C_{i,k}$, which is defined in Cartesian representation as

$$C_{i,k} = \frac{1}{N_R} Tr(MP_i(a)P_k(A)M^\dagger), \tag{3.47}$$

where $P_i(a)$ and $P_k(A)$ are the vector polarization operators for the particle a and the target A, respectively. For the general case, the tensor polarizations can be treated for a and/or A. These coefficients are expected to give information on $a - A$ spin-dependent interactions.

Chapter 4

Invariant Amplitude Method and Scattering of Spin $\frac{1}{2}$, 1 and $\frac{3}{2}$ Particles

As discussed in previous parts of this book, polarization phenomena in reactions are expected to provide information on spin-dependent interactions between the particles concerned. However, a definite method of connecting polarization observables to the interactions has not been presented so far. In this chapter, we will propose the invariant amplitude method as a prescription to solve such problems [10]. The presented theory is partially similar to the one in ref. [11] but is, at the same time, more general than that in the reference.

Theoretically, polarization observables are calculated by transition amplitudes of the reaction, which are provided by the matrix elements of the transition matrix constructed of the interactions. On the other hand, the interaction is classified according to the tensorial property in the spin space; for example central interactions are classified as scalars and spin-orbit interactions are classified as vectors in the spin space. This means that the transition amplitude can be decomposed into tensor terms in the spin space, each of which represents contribution of the corresponding spin-dependent interaction. Such treatments will be helpful in identifying the contribution of a particular spin-dependent interaction in the calculated observables.

4.1   Decomposition of Transition Amplitudes into Invariant Amplitudes

Let us consider a reaction $a + A \rightarrow b + B$, denoting the transition matrix of the reaction by M. The transition amplitude is given by the matrix element of M between the initial and final states at infinity. These states are designated by four sets of quantities: the aA and bB relative momenta, $\boldsymbol{k}_i$ and $\boldsymbol{k}_f$, the z-components ν_a and ν_A of the spins $\boldsymbol{s}_a$ and $\boldsymbol{s}_A$, and ν_b and ν_B of the spins $\boldsymbol{s}_p$ and $\boldsymbol{s}_B$.

Next, we will expand M into the spin space tensor s^K_κ where K and κ stand for the rank and the z-component, respectively, so that

$$M = \sum_K M^{(K)} \quad \text{with} \quad M^{(K)} = \sum_\kappa (-)^\kappa s^K_{-\kappa} R_{K\kappa}, \qquad (4.1)$$

where $R_{K\kappa}$ is the coordinate space tensor that guarantees the scalar nature of $M^{(K)}$, together with $s^K_{-\kappa}$. The matrix element of $M^{(K)}$ gives the K-rank component of the transition amplitude, which will be calculated below.

The matrix element of $s^K_{-\kappa}$ is evaluated by the Wigner–Eckart theorem [3], by introducing the channel spins $\boldsymbol{s}_i$ ($= \boldsymbol{s}_a + \boldsymbol{s}_A$) and $\boldsymbol{s}_f$ ($= \boldsymbol{s}_b + \boldsymbol{s}_B$) for the initial and final states

$$\langle \nu_b \nu_B | s^K_{-\kappa} | \nu_a \nu_A \rangle = \sum_{s_i s_f} (s_a s_A \nu_a \nu_A \mid s_i \nu_i)(s_b s_B \nu_b \nu_B \mid s_f \nu_f)$$

$$\times \frac{1}{\sqrt{2s_f + 1}}(s_i K \nu_i - \kappa \mid s_f \nu_f)(s_f \| s^K \| s_i),$$

$$(4.2)$$

where $(s_f \| s^K \| s_i)$ is the physical part of the matrix element $\langle s_f \nu_f | s^K_{-\kappa} | s_i \nu_i \rangle$.

Since the matrix elements of $R_{K\kappa}$ include only $\boldsymbol{k}_i$ and $\boldsymbol{k}_f$ as available variables, the geometrical part of the matrix elements is described in terms of the tensor produced by $\boldsymbol{k}_i$ and $\boldsymbol{k}_f$ such as

$$[C'_p(\hat{k}_i) \otimes C_{p'}(\hat{k}_f)]^K_\kappa,$$

where $\hat{k}_i$ and $\hat{k}_f$ are the angular variables of $\mathbf{k}_i$ and $\mathbf{k}_f$, respectively and

$$C_{lm}(\theta, \phi) = \sqrt{\frac{4\pi}{2l+1}} Y_l^m(\theta, \phi), \qquad (4.3)$$

where Y_l^m are the spherical harmonics [3].

Combining the above considerations on the matrix elements of $s_{-\kappa}^K$ and $R_{K\kappa}$ and utilizing

$$(s_i K \nu_i - \kappa \mid s_f \nu_f) = (-)^{s_i - \nu_i} \sqrt{\frac{2s_f + 1}{2K + 1}} (s_i s_f \nu_i - \nu_f \mid k\kappa), \qquad (4.4)$$

one can write the result, [10],

$$\langle \nu_b \nu_B; \mathbf{k}_f | M^{(K)} | \nu_a \nu_A; \mathbf{k}_i \rangle$$

$$= \sum_{s_i s_f} (s_a s_A \nu_a \nu_A \mid s_i \nu_i)(s_b s_B \nu_b \nu_B \mid s_f \nu_f)(-)^{s_f - \nu_f}$$

$$\times (s_i s_f \nu_i - \nu_f \mid K\kappa) \sum_{l_i = \bar{K} - K}^{K} [C_{l_i}(\hat{k}_i) \otimes C_{l_f = \bar{K} - l_i}(\hat{k}_f)]_\kappa^K$$

$$\times F(s_i s_f K l_i; E \cos \theta). \qquad (4.5)$$

The physical parts of the matrix elements and some phase factors are included in $F(s_i s_f k l_i; E \cos \theta)$, which also depends on the $a - A$ relative energy E and the relative angle θ between $\mathbf{k}_i$ and $\mathbf{k}_f$. Since the properties for the rotations of the coordinate axes are already included in the geometrical factors like CG coefficients and $[C_{l_i}; \otimes C_{l_f}]_\kappa^K$, $F(s_i s_f K l_i; E \cos \theta)$ is now invariant under the rotation and hereafter will be referred to as the invariant amplitude. The parameter $\bar{K}$ is chosen as follows. When the reaction conserves the total parity of the particles, $\bar{K} = K$ for $K =$ even and $\bar{K} = K + 1$ for $K =$ odd. When the reaction changes the parity, $\bar{K} = K + 1$ for $k =$ even and $\bar{K} = K$ for $K =$ odd.

The choice of l_i and l_f in Eq. (4.5) is based on the reduction formula of the tensor rank of C_p, that is

$$(p - 1100 \mid p0)C_p = [C_{p-1} \otimes C_1]^p. \tag{4.6}$$

When we apply this to both $C_p(\hat{k}_i)$ and $C_{p'}(\hat{k}_f)$, $[C_p(\hat{k}_i) \otimes C_{p'}(\hat{k}_f)]^K$ is decomposed to terms of $[C_{p-1}(\hat{k}_i) \otimes C_{p'-1}(\hat{k}_f)]^K$, $[C_{p-2}(\hat{k}_i) \otimes C_{p'}(\hat{k}_f)]^K$, $[C_p(\hat{k}_i) \otimes C_{p'-2}(\hat{k}_f)]$ and $[C_{p-2}(\hat{k}_i) \otimes C_{p'-2}(\hat{k}_f)]^K$. After repeating such reductions, finally we will get the formula by the tensors of the lowest rank,

$$[C_p(\hat{k}_i) \otimes C_{p'}(\hat{k}_f)]^K = \sum_{q=\bar{K}-K}^{K} A(pp'qK; \cos\theta)[C_q(\hat{k}_i) \otimes C_{\bar{K}-q}(\hat{k}_f)]^K,$$

$$(4.7)$$

where $A(pp'qK; \cos\theta)$ can be obtained by a recurrence formula. However, since this factor can be absorbed in $F(s_i s_f K l_i; E \cos\theta)$, the explicit form is not necessary. To test the validity of the above derivations, we will apply Eq. (4.5) to the case with no parity change. Up to $K = 2$, for example, Eq. (4.5) gives

$$l_i = l_f = 0 \quad \text{for } K = 0,$$
$$l_i = l_f = 1 \quad \text{for } K = 1,$$
and $\quad l_i = 2, \ l_f = 0, \ l_i = l_f = 1, \ l_i = 0, \ l_f = 2 \quad \text{for } K = 2.$

These agree with what we consider to be the lowest rank tensors for given K's.

Since $F(s_i s_f K l_i; E \cos\theta)$ represents the contributions of the central interaction, the spin-orbit one and the tensor one by the terms of $K = 0$, 1 and 2, respectively, one can identify contributions of these interactions in calculations and also in the analyses of experimental data when the observables are described in terms of the invariant amplitudes. Further, it will be noticed that the decomposition of the transition matrix is quite general, and hence the higher orders of the interactions are automatically included according to their tensor ranks. In this sense, the interaction considered at present will be understood to be effective interactions. Also, when the projectile and/or target are deformed, the effect of their deformation will appear as spin-dependent interactions in accordance with their tensorial properties. For example, in the deuteron case, the D-state effect will produce a second-rank tensor interaction. These effects will be further examined in the following sections.

As mentioned already, explicit calculations of the invariant amplitude F need some reaction models. Otherwise, we will deal with F as a phenomenological parameter. For that purpose, it will be convenient to expand F in terms of the Legendre function $P_l(\cos\theta)$ to describe the θ dependence. That is, discarding E from the variables,

$$F(s_i s_f K l_i, \cos\theta) = F_0(s_i s_f K l_i) \sum_l \gamma_l(s_i s_f K l_i) P_l(\cos\theta), \qquad (4.8)$$

where F_0 and γ_l are treated as parameters to be fixed so as to reproduce experimental data. In this sense this expansion has a similar nature to the one used in phase shift analyses of elastic scattering amplitudes. That is, at low energies, a few partial waves are effective, as in Eq. (4.8).

Finally, it will be emphasized that, in elastic scattering, the time reversal theorem induces an important effect on the scattering amplitudes. Let us consider elastic scattering of a particle a by a spinless target A. The scattering amplitudes equivalent to those for the time-reversed scattering. That is.

$$\langle \nu_f, \boldsymbol{k}_f | M^{(K)} | \nu_i, \boldsymbol{k}_i \rangle = \langle -\nu_i, -\boldsymbol{k}_i | M^{(K)} | -\nu_f, -\boldsymbol{k}_f \rangle, \qquad (4.9)$$

which can be written by the use of the invariant amplitudes as

$$F(s_a s_a K l_i; E \cos\theta) = F(s_a s_a K \bar{K} - l_i; E \cos\theta). \qquad (4.10)$$

This restricts the freedom of the scattering amplitudes for $s_a \geq 1$. In the subsequent sections, we will treat scattering of particles by spinless targets as applications of the invariant amplitude method, in which the restriction Eq. (4.10) is taken into account explicitly in analyses of the scattering.

4.2 Elastic Scattering of $s = \frac{1}{2}$ Particle

Elastic scattering of $s = \frac{1}{2}$ particles by spinless targets is one of the most popular nuclear events. For example, the scattering of nucleons by nuclei has been intensively investigated via both experiments and calculations. Since the nucleus is treated as one body, the nucleon–nucleus interaction consists of the central interaction and the spin-orbit (LS) interaction, among which our interests have been focused on the LS interaction. In the following, we will investigate the scattering amplitudes by applying the invariant amplitude method to clarify how to get information on the LS interaction.

Let us describe the scattering matrix M by

$$M = \begin{pmatrix} A & B \\ -B & A \end{pmatrix}, \tag{4.11}$$

where the rows and columns express the initial and final states, and are designated by the eigenvalue of the z-component of the spin of the particle as $s_z = \frac{1}{2}, -\frac{1}{2}$ from left to right and from top to bottom, respectively. The phase of the wave function is chosen such that $T\psi_{s\nu} = (-)^{s+\nu}\psi_{s-\nu}$, with T being the time reversal operator. Choosing the same coordinate axes as those in Fig. 3.1, $z \parallel \boldsymbol{k}_i$ and $y \parallel \boldsymbol{k}_i \times \boldsymbol{k}_y$, we will apply Eq. (4.5) to the scattering amplitudes A and B. Discarding E from the variable of the amplitude F,

$$A = \sqrt{\frac{1}{2}} C_{00} F\left(\frac{1}{2} \, \frac{1}{2} \, 0 \, 0, \cos\theta\right), \tag{4.12}$$

$$B = -\sqrt{\frac{1}{2}} C_{11} F\left(\frac{1}{2} \, \frac{1}{2} \, 1 \, 1, \cos\theta\right), \tag{4.13}$$

where

$$C_{00}(\theta, 0) = 1 \tag{4.14}$$

and

$$C_{1\pm1}(\theta, 0) = \mp\sqrt{\frac{1}{2}} \sin\theta. \tag{4.15}$$

Equation (4.12) shows that the amplitude A is a scalar in the spin space, which is induced by the central interaction. Similarly,

Eq. (4.13) shows the amplitude B to be governed by the spin vector interaction, which is interpreted as an LS interaction. The above features of A and B will be reflected on the scattering observables. The differential cross section $\frac{d\sigma}{d\Omega}$ given by Eq. (3.8) is now described as

$$\frac{d\sigma}{d\Omega}\bigg|_0 = \left(\frac{\mu_a A}{2\pi}\right)^2 \frac{1}{2} N_R \qquad (4.16)$$

with

$$N_R = 2\{|A|^2 + |B|^2\}$$

$$= \left| F\left(\frac{1}{2} \frac{1}{2} \, 0 \, 0, \cos\theta\right) \right|^2 + C_{11}^2 \left| F\left(\frac{1}{2} \frac{1}{2} \, 1 \, 1, \cos\theta\right) \right|^2. \qquad (4.17)$$

When the LS interaction is weak compared to the central one, that is, $|F(\frac{1}{2} \frac{1}{2} \, 1 \, 1, \cos\theta)|^2 \ll |F(\frac{1}{2} \frac{1}{2} \, 0 \, 0, \cos\theta)|^2$, the differential cross section receives dominant contributions from the central interaction.

The vector polarization of the scattered particle p_y is given by Eqs. (3.13) and (3.15) as

$$p_y = \sqrt{2} \, it_{11} = \frac{4}{N_R} \text{Im}(AB^*)$$

$$= -\frac{2}{N_R} C_{11} \, \text{Im} \left\{ F\left(\frac{1}{2} \frac{1}{2} \, 0 \, 0, \cos\theta\right) F^* \left(\frac{1}{2} \frac{1}{2} \, 1 \, 1, \cos\theta\right) \right\}, \qquad (4.18)$$

and the polarization transfer coefficient K_x^z is given by Eq. (3.44),

$$K_x^z = -K_z^x = \frac{4}{N_R} \text{Re}(AB^*)$$

$$= -\frac{2}{N_R} C_{11} \, \text{Re} \left\{ F\left(\frac{1}{2} \frac{1}{2} \, 0 \, 0, \cos\theta\right) F^* \left(\frac{1}{2} \frac{1}{2} \, 1 \, 1, \cos\theta\right) \right\}. \qquad (4.19)$$

Thus, one can get the full information of two amplitudes $F(\frac{1}{2} \frac{1}{2} \, 0 \, 0, \cos\theta)$ and $F(\frac{1}{2} \frac{1}{2} \, 1 \, 1, \cos\theta)$, their magnitudes and their relative phase, from $\frac{d\sigma}{d\Omega}$, p_y and K_x^z. This gives information on the strength of the LS interaction as well as that of the central interaction. Additional information on these interactions is obtained from the combination of $\frac{d\sigma}{d\Omega}$ and K_z^x $(= K_z^z)$,

$$K_x^x = K_z^z = \frac{2}{N_R}(|A|^2 - |B|^2)$$

$$= \frac{1}{N_R}\left\{\left|F\left(\frac{1}{2}\,\frac{1}{2}\,0\,0,\cos\theta\right)\right|^2 - C_{11}^2\left|F\left(\frac{1}{2}\,\frac{1}{2}\,1\,1,\cos\theta\right)\right|^2\right\}.$$

$$(4.20)$$

Quantitative analyses will be given for elastic scattering of protons in the next chapter.

4.3 Elastic Scattering of $s = 1$ Particle

As shown by Eq. (2.9), the spin density of a particle with $s = 1$, for instance a deuteron, consists of the scalar, vector and 2nd-rank tensor components. This means that the polarization of the $s = 1$ particle is described by the vector and the second-rank tensors in the spin space. In the following, we will describe such polarization observables by the invariant amplitudes, which help clarify the contributions of spin-dependent interactions. As a typical case, we will study the elastic scattering of the $s = 1$ particle by a spinless nucleus.

The transition matrix of the scattering, M, is expressed by the use of the matrix elements as in ref. [12],

$$M = \begin{pmatrix} A & B & C \\ D & E & -D \\ C & -B & A \end{pmatrix}, \qquad (4.21)$$

where the rows and columns express the initial and final states, respectively. The matrix elements are designated by the eigenvalues of the z component of the particle spin, $s_z = 1, 0, -1$, from the left to the right for the initial state, and from the top to the bottom for the final state. The phase of the wave function is chosen similarly to that of the previous section. The coordinate axes are chosen as $z \parallel \mathbf{k}_i$

and $y \parallel k_i \times k_f$, where k_i and k_f are the initial and final relative momenta.

Applying Eq. (4.5) to the matrix elements $A \sim E$, we get

$$A = \sqrt{\frac{1}{3}}C_{00}F(1\ 1\ 0\ 0, \cos\theta) + \sqrt{\frac{1}{6}}C_{20}F(1\ 1\ 2\ 0, \cos\theta)$$

$$+ \frac{1}{3}C_{10}F(1\ 1\ 2\ 1, \cos\theta) + \sqrt{\frac{1}{6}}C_{00}F(1\ 1\ 2\ 2, \cos\theta), \quad (4.22)$$

$$E = \sqrt{\frac{1}{3}}C_{00}F(1\ 1\ 0\ 0, \cos\theta) - \sqrt{\frac{2}{3}}C_{20}F(1\ 1\ 2\ 0, \cos\theta)$$

$$- \frac{2}{3}C_{10}F(1\ 1\ 2\ 1, \cos\theta) - \sqrt{\frac{2}{3}}C_{00}F(1\ 1\ 2\ 2, \cos\theta), \quad (4.23)$$

$$B = -\frac{1}{2}C_{11}F(1\ 1\ 1\ 1, \cos\theta) - \sqrt{\frac{1}{2}}C_{21}F(1\ 1\ 2\ 0, \cos\theta)$$

$$- \frac{1}{2}C_{11}F(1\ 1\ 2\ 1, \cos\theta), \quad (4.24)$$

$$D = \frac{1}{2}C_{11}F(1\ 1\ 1\ 1, \cos\theta) - \sqrt{\frac{1}{2}}C_{21}F(1\ 1\ 2\ 0, \cos\theta)$$

$$- \frac{1}{2}C_{11}F(1\ 1\ 2\ 1, \cos\theta), \quad (4.25)$$

$$C = C_{22}F(1\ 1\ 2\ 0, \cos\theta), \quad (4.26)$$

where

$$C_{10} = \cos\theta, \quad (4.27)$$

$$C_{20} = \frac{1}{2}(3\cos^2\theta - 1), \quad C_{21} = -\sqrt{\frac{3}{2}}\cos\theta\sin\theta, \quad C_{22} = \sqrt{\frac{3}{8}}\sin^2\theta$$

$$(4.28)$$

and other C's are the same as those in the previous section.

Because the inverse process is equivalent to that of the original, we have from Eq. (4.10)

$$F(1\ 1\ 2\ 2, \cos\theta) = F(1\ 1\ 2\ 0, \cos\theta), \quad (4.29)$$

which gives

$$C = A - E - \sqrt{2}(B + D)\cot\theta. \tag{4.30}$$

Since the freedom of the tensor amplitudes are restricted by the above relation, one can choose two tensor amplitudes to be independent. Considering this condition, we adopt the following amplitudes for the scattering:

$$U = 2A + E = \sqrt{3}F(1\ 1\ 0\ 0, \cos\theta), \tag{4.31}$$

$$S = B - D = \frac{1}{\sqrt{2}}\sin\theta\ F(1\ 1\ 1\ 1, \cos\theta), \tag{4.32}$$

$$T_\alpha = B + D = \frac{1}{\sqrt{2}}\sin\theta\{\sqrt{6}\cos\theta\ F(1\ 1\ 2\ 0, \cos\theta) + F(1\ 1\ 2\ 1, \cos\theta)\} \tag{4.33}$$

and

$$T_\beta = C + \frac{1}{2\sqrt{2}}(B + D)\cot\frac{\theta}{2}$$

$$= \frac{1}{2}\cos^2\frac{\theta}{2}\{\sqrt{6}F(1\ 1\ 2\ 0, \cos\theta) + F(1\ 1\ 2\ 1, \cos\theta)\}, \tag{4.34}$$

where U is the scalar amplitude, S is the vector amplitude and T_α and T_β are the tensor amplitudes.

Earlier, three types of tensor interactions have been suggested for the deuteron–nucleus interaction.

$$T_R = (S_2(\boldsymbol{s}, \boldsymbol{s}) \cdot R_2(\boldsymbol{R}, \boldsymbol{R}))V_R(R), \tag{4.35}$$

$$T_L = (S_2(\boldsymbol{s}, \boldsymbol{s}) \cdot R_2(\boldsymbol{L}, \boldsymbol{L}))V_L(R), \tag{4.36}$$

$$T_P = (S_2(\boldsymbol{s}, \boldsymbol{s}) \cdot R_2(\boldsymbol{P}, \boldsymbol{P}))V_P(R), \tag{4.37}$$

where $\boldsymbol{s}$ is the deuteron spin and $\boldsymbol{R}$, $\boldsymbol{L}$ and $\boldsymbol{P}$ are the deuteron–nucleus relative coordinate, relative angular momentum and relative momentum respectively. S_2 and R_2 represent the second-rank tensors, in the spin space and the coordinate space, i.e.

$$S_{2\mu}(\boldsymbol{s}, \boldsymbol{s}) = [\boldsymbol{s} \otimes \boldsymbol{s}]_\mu^2 \tag{4.38}$$

and

$$R_{2\mu}(\boldsymbol{O},\boldsymbol{O}) = [\boldsymbol{O} \otimes \boldsymbol{O}]^2_{\mu} \quad \text{with } \boldsymbol{O} = \boldsymbol{R}, \boldsymbol{L} \quad \text{or} \quad \boldsymbol{P}. \tag{4.39}$$

As will be shown in Chap. 6, in the case of the deuteron, the T_R type tensor interaction is induced by the D-state admixture in the ground state and the T_L-type tensor interaction is produced by the deuteron–nucleus LS interaction in the second-order perturbation treatment. Then we will consider such T_R type and T_L type tensor interactions for the deuteron scattering, neglecting the T_P type tensor one. Further, it is shown in the Appendix that the amplitude T_α represents the scattering by the T_R-type tensor interaction and the amplitude T_β represents the scattering by the T_L-type tensor interaction.

Using Eqs. (4.31)–(4.34), one can express $A, \ldots, E$ in terms of $U, \ldots, T_\beta$, as

$$A = \frac{1}{3}\left(U + T_\beta + \frac{1}{2\sqrt{2}}\frac{3\cos\theta - 1}{\sin\theta}T_\alpha \right), \tag{4.40}$$

$$E = \frac{1}{3}\left(U - 2T_\beta - \frac{1}{\sqrt{2}}\frac{3\cos\theta - 1}{\sin\theta}T_\alpha \right), \tag{4.41}$$

$$B = \frac{1}{2}(T_\alpha + S), \tag{4.42}$$

$$D = \frac{1}{2}(T_\alpha - S), \tag{4.43}$$

and

$$C = T_\beta - \frac{1}{2\sqrt{2}}T_\alpha \cot\frac{\theta}{2}. \tag{4.44}$$

Scattering observables, for instance of the deuteron, are described by $A, \ldots, E$, and then we get them in terms of $U, \ldots, T_\beta$. As examples we will give the differential cross section $\frac{d\sigma}{d\Omega}\big|_0$, the vector analyzing power A_y and the tensor analyzing powers A_{xx}, A_{yy} and A_{xz}.

$$\frac{d\sigma}{d\Omega}\bigg|_0 = \left(\frac{\mu_a A}{2\pi}\right)^2 \frac{1}{3}N_R \tag{4.45}$$

with

$$N_R = 2(|A|^2 + |B|^2 + |C|^2 + |D|^2) + |E|^2$$

$$= \frac{1}{3}\left\{|U|^2 + 3|S|^2 + 8|T_\beta|^2 + \frac{4}{\sin^2\theta}|T_\alpha|^2 - \frac{4\sqrt{2}}{\sin\theta}\text{Re}(T_\alpha^* T_\beta)\right\},$$

$$(4.46)$$

$$A_y = \frac{2\sqrt{2}}{N_R}\text{Im}(AB^* - ED^* + BC^*)$$

$$= \frac{2\sqrt{2}}{3N_R}\text{Im}\left\{\left(U - 2T_\beta + \frac{1}{\sqrt{2}\sin\theta}T_\alpha\right)S^*\right\}, \qquad (4.47)$$

$$A_{xx} = \frac{1}{N_R}\{-|A|^2 + 2|B|^2 - |C|^2 - 4|D|^2 + |E|^2 + 6\,\text{Re}(A^*C)\}$$

$$= \frac{1}{N_R}\left\{\frac{4}{3}\text{Re}(UT_\beta^*) - \sqrt{2}\frac{1+3\cos\theta}{3\sin\theta}\text{Re}(UT_\alpha^*) - \frac{1+3\cos\theta}{3\sin^2\theta}|T_\alpha|^2\right.$$

$$\left. + \frac{4}{3}|T_\beta|^2 - \frac{1}{2}|S|^2 - 2\sqrt{2}\frac{1-3\cos\theta}{3\sin\theta}\text{Re}(T_\alpha T_\beta^*) + 3\,\text{Re}(ST_\alpha^*)\right\},$$

$$(4.48)$$

$$A_{yy} = \frac{1}{N_R}\{-|A|^2 + 2|B|^2 - |C|^2 + 2|D|^2 + |E|^2 - 6\,\text{Re}(A^*C)\}$$

$$= -\frac{2}{N_R}\left\{\frac{4}{3}\text{Re}(UT_\beta^*) - \frac{\sqrt{2}}{3\sin\theta}\text{Re}(UT_\alpha^*) - \frac{1}{3\sin^2\theta}|T_\alpha|^2\right.$$

$$\left. + \frac{4}{3}|T_\beta|^2 - \frac{1}{2}|S|^2 - \frac{2\sqrt{2}}{3\sin\theta}\text{Re}(T_\alpha T_\beta^*)\right\}, \qquad (4.49)$$

and

$$A_{xz} = \frac{3\sqrt{2}}{N_R}\text{Re}(AB^* + DE^* - CB^*)$$

$$= \frac{\sqrt{2}}{N_R}\text{Re}\left\{\left(U + \frac{3}{\sqrt{2}}S\cot\theta - 2T_\beta + \frac{1}{\sqrt{2}\sin\theta}T_\alpha\right)T_\alpha^*\right\}.$$

$$(4.50)$$

To examine the contributions of the spin-dependent interactions in a simple way, we will introduce an approximation of weak spin-dependent interactions (hereafter abbreviated by WSDI), which neglects the second-order terms of the spin-dependent amplitudes, S, T_α, and T_β, where we assume the contributions to be small. Then we get

$$\left.\frac{d\sigma}{d\Omega}\right|_0 = \left(\frac{\mu_{aA}}{2\pi}\right)^2 \frac{1}{3} N_R \quad \text{with } N_R = \frac{1}{3}|U|^2, \tag{4.51}$$

$$A_y = \frac{2\sqrt{2}}{3N_R} \text{Im}(US^*), \tag{4.52}$$

$$A_{xx} = \frac{1}{N_R}\left\{\frac{4}{3}\text{Re}(UT_\beta^*) - \sqrt{2}\frac{1+3\cos\theta}{3\sin\theta}\text{Re}(UT_\alpha^*)\right\}, \tag{4.53}$$

$$A_{yy} = -\frac{2}{N_R}\left\{\frac{4}{3}\text{Re}(UT_\beta^*) - \sqrt{2}\frac{1}{3\sin\theta}\text{Re}(UT_\alpha^*)\right\}, \tag{4.54}$$

$$A_{xz} = \frac{\sqrt{2}}{N_R}\text{Re}(UT_\alpha^*). \tag{4.55}$$

From Eq. (4.51) we see that the differential cross section is governed by central interactions. Equations (4.52) and (4.55) show that the analyzing powers A_y and A_{xz} are proportional to S and T_α, respectively. Then, A_y gives a measure of the strength of the LS interaction and A_{xz} gives a measure of the strength of the T_R tensor interaction. Further, one can see, from the combination of Eqs. (4.53) and (4.54), the contributions of two tensor amplitudes, T_α and T_β in a separate form. When T_β is dominant and T_α is negligibly small, A_{xx} and A_{yy} satisfy

$$A_{yy}/A_{zx} \simeq -2, \tag{4.56a}$$

while when T_α is dominant

$$A_{yy}/A_{xx} \simeq -2/(1+3\cos\theta). \tag{4.56b}$$

As shown in Fig. 4.1, the relation (4.56a) explains the characteristics observed in the relative sign and magnitude of the measured A_{xx} to those of the measured A_{yy} at most angles, in the scattering by the

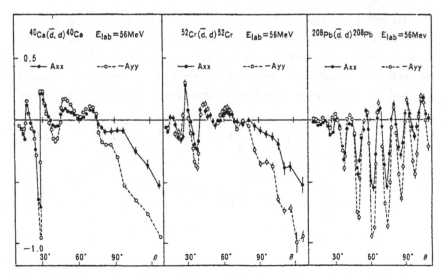

Fig. 4.1   Comparison between A_{xx} and A_{yy} in elastic scattering of deuterons by ^{40}Ca, ^{52}Cr and ^{208}Pb nuclei at $E_d = 56$ MeV.

^{40}Ca, ^{52}Cr and ^{208}Pb targets at $E_d = 56$ MeV. On the other hand, Eq. (4.56b) gives $A_{yy}/A_{xx} = -2$ at $\theta = 90°$ but $A_{yy}/A_{xx} = +4$ at $\theta = 120°$. Such drastic variations in the θ-dependence of A_{yy}/A_{xx} cannot explain the experimental data. These results suggest that the greatest part of such analyzing powers comes from the T_L-type tensor interaction, which is interpreted as a second-order effect of the LS interaction.

As previously discussed, the amplitude T_α is produced by the deuteron D state. To investigate such D-state effects by the analyzing powers, the linear combination $2A_{xx} + A_{yy}$ will be useful. From Eqs. (4.48) and (4.49),

$$2A_{xx} + A_{yy}$$

$$= \frac{2}{N_R}\mathrm{Re}\left\{\left(-\sqrt{2}\cot\theta\, U - \frac{\cot\theta}{\sin\theta}T_\alpha + 2\sqrt{2}\cot\theta\, T_\beta + 3S\right)^* T_\alpha\right\},$$
$$(4.57)$$

which is sensitive to T_α because the right-hand side is approximately proportional to T_α. Further, $X_2 \equiv -\sqrt{\frac{1}{3}}(2A_{xx} + A_{yy})$ is equivalent to $\sqrt{\frac{3}{2}}T_{20} - T_{22}$ in the spherical representation, which is well-known as the measure of the D-state effect.

4.4 Elastic and Inelastic Scattering of $s = \frac{3}{2}$ Particle

In this section, we will study elastic scattering of $s = \frac{3}{2}$ particles by spinless nuclei. In Chap. 7, as an application, scattering of ^{7}Li will be investigated in comparison with experimental data. The data include those of the inelastic scattering to the $\frac{1}{2}^-$ excited state of ^{7}Li in addition to those of the elastic scattering. Then, both scatterings are treated in this section.

In the elastic scattering, as is shown in Eq. (4.5), the $s = \frac{3}{2}$ particle produces scattering amplitudes of the scalar, vector and second-rank and third-rank tensors in the spin space. This suggests the participation of LS interactions as well as second-rank and third-rank tensor interactions in the spin observables. These will be investigated by the invariant amplitude method.

Let us denote the transition matrix of the elastic scattering by M, which is represented by the matrix elements as

$$M = \begin{pmatrix} A & B & C & D \\ E & F & G & H \\ H & -G & F & -E \\ -D & C & -B & A \end{pmatrix}, \tag{4.58}$$

where the rows and columns express the initial and final states, respectively. They are designated by the eigenvalue of the z-component of the particle spin, $s_z = \frac{3}{2}, \frac{1}{2}, -\frac{1}{2}, -\frac{3}{2}$ from the left to the right in the former and from the top to the bottom in the latter. The phase of the wave function is similar to that chosen in Sec. 4.2. Choosing the coordinate system as $z \parallel \boldsymbol{k}_i$ and $y \parallel \boldsymbol{k}_i \times \boldsymbol{k}_y$, one can describe $A \sim H$ in terms of the invariant amplitudes $F(s_a s_a k l_i, \cos\theta)$ by the use of Eq. (4.5). Further, the time reversal invariance Eq. (4.10) gives

$$F\left(\frac{3}{2}\frac{3}{2}\,2\,2, \cos\theta\right) = F\left(\frac{3}{2}\frac{3}{2}\,2\,0, \cos\theta\right) \tag{4.59}$$

and

$$F\left(\frac{3}{2}\frac{3}{2}\ 3\ 3,\cos\theta\right) = F\left(\frac{3}{2}\frac{3}{2}\ 3\ 1,\cos\theta\right), \qquad (4.60)$$

which produce

$$C + H = \sqrt{3}(A - F) - 2(B + E)\cot\theta \qquad (4.61)$$

and

$$B - E = \sqrt{3}(G + D) + 2(C - H)\cot\theta. \qquad (4.62)$$

These restrict the freedom of the second-rank and third-rank tensor amplitudes.

Considering the above restrictions, we will define the amplitudes of the scalar U, the vector S, the second-rank tensor $T_{2\alpha}$ and $T_{2\beta}$ and the third-rank tensor $T_{3\alpha}$ and $T_{3\beta}$, as

$$U \equiv A + F = F\left(\frac{3}{2}\frac{3}{2}\ 0\ 0,\cos\theta\right), \qquad (4.63)$$

$$S \equiv G + \sqrt{\frac{3}{4}}(B - E) = \sqrt{\frac{5}{8}}\sin\theta\, F\left(\frac{3}{2}\frac{3}{2}\ 1\ 1,\cos\theta\right), \qquad (4.64)$$

$$T_{2\alpha} \equiv A - F = \frac{1}{2}(3\cos^2\theta + 1)F\left(\frac{3}{2}\frac{3}{2}\ 2\ 0,\cos\theta\right)$$
$$+ \sqrt{\frac{2}{3}}\cos\theta\, F\left(\frac{3}{2}\frac{3}{2}\ 2\ 1,\cos\theta\right), \qquad (4.65)$$

$$T_{2\beta} \equiv C + H = \sqrt{\frac{3}{4}}\sin^2\theta\, F\left(\frac{3}{2}\frac{3}{2}\ 2\ 0,\cos\theta\right), \qquad (4.66)$$

$$T_{3\alpha} \equiv C - H = \sqrt{\frac{5}{4}}\sin^2\theta\cos\theta\, F\left(\frac{3}{2}\frac{3}{2}\ 3\ 1,\cos\theta\right)$$
$$+ \sqrt{\frac{3}{8}}\sin^2\theta\, F\left(\frac{3}{2}\frac{3}{2}\ 3\ 2,\cos\theta\right), \qquad (4.67)$$

$$T_{3\beta} \equiv D = \sqrt{\frac{15}{64}}\sin^3\theta\, F\left(\frac{3}{2}\frac{3}{2}\ 3\ 1,\cos\theta\right). \qquad (4.68)$$

Using the above amplitudes, we get $A \sim H$ as

$$A = \frac{1}{2}(U + T_{2\alpha}), \tag{4.69}$$

$$B = \frac{1}{4}(\sqrt{3}T_{2\alpha} - T_{2\beta})\tan\theta + \frac{\sqrt{3}}{5}(S + T_{3\beta}) + \frac{2}{5}T_{3\alpha}\cot\theta, \tag{4.70}$$

$$C = \frac{1}{2}(T_{2\beta} + T_{3\alpha}), \tag{4.71}$$

$$D = T_{3\beta}, \tag{4.72}$$

$$E = \frac{1}{4}(\sqrt{3}T_{2\alpha} - T_{2\beta})\tan\theta - \frac{\sqrt{3}}{5}(S + T_{3\beta}) - \frac{2}{5}T_{3\alpha}\cot\theta, \tag{4.73}$$

$$F = \frac{1}{2}(U - T_{2\alpha}), \tag{4.74}$$

$$G = \frac{2}{5}S - \frac{2\sqrt{3}}{6}T_{3\alpha}\cot\theta - \frac{3}{5}T_{3\beta}, \tag{4.75}$$

$$H = \frac{1}{2}(T_{2\beta} - T_{3\alpha}). \tag{4.76}$$

Next, we will give some observables in terms of the scattering amplitudes, for which experimental data are available [14]. In the representation, we will keep the third-rank tensor amplitudes $T_{3\alpha}$ and $T_{3\beta}$ only when they are accompanied by the scalar amplitude U, which describes the central interaction, because of the weakness of the third-rank tensor interaction.

The differential cross section $\dfrac{d\sigma}{d\Omega}$ is given as

$$\frac{d\sigma}{d\Omega} = \left(\frac{\mu_{aA}}{2\pi}\right)^2 \frac{1}{4}N_R \tag{4.77}$$

with

$$N_R = 2(|A|^2 + |B|^2 + |C|^2 + |D|^2 + |E|^2 + |F|^2 + |G|^2 + |H|^2)$$
$$= |U|^2 + |T_{2\alpha}|^2 + |T_{2\beta}|^2 + \frac{4}{5}|S|^2 + \frac{1}{4}|T|^2, \tag{4.78}$$

where

$$T \equiv (\sqrt{3}T_{2\alpha} - T_{2\beta})\tan\theta. \tag{4.79}$$

The vector analyzing power iT_{11} and the tensor analyzing powers T_{20}, T_{21} and T_{22} are given as

$$iT_{11} = -\frac{2\sqrt{2}}{\sqrt{5}N_R}\text{Im}\{\sqrt{3}(A^*B + C^*D + E^*F + G^*H)$$

$$+ 2(B^*C + F^*G)\}$$

$$= \frac{2\sqrt{2}}{\sqrt{5}N_R}\text{Im}\left\{US^* + \frac{2}{5}ST_{2\alpha}^* + \frac{2\sqrt{3}}{5}ST_{2\beta}^*\right\}, \tag{4.80}$$

$$T_{20} = \frac{2}{N_R}(|A|^2 + |B|^2 + |C|^2 + |D|^2 + |E|^2 - |F|^2 - |G|^2 + |H|^2)$$

$$= \frac{2}{N_R}\left(\text{Re}\left\{U^*T_{2\alpha} - \frac{\sqrt{3}}{5}S^*T\right\} - \frac{4}{25}|S|^2\right), \tag{4.81}$$

$$T_{21} = -\frac{2\sqrt{2}}{N_R}\text{Re}\{A^*B - C^*D + E^*F - G^*H\}$$

$$= -\frac{2\sqrt{2}}{N_R}\text{Re}\left\{\frac{1}{4}UT^* + \frac{1}{5}ST^*\cot\theta\right\}, \tag{4.82}$$

$$T_{22} = \frac{2\sqrt{2}}{N_R}\text{Re}\{A^*C + B^*D + E^*G + F^*H\}$$

$$= \frac{\sqrt{2}}{N_R}\left(\text{Re}\left\{UT_{2\beta}^* + \frac{1}{5}(S^*T)\right\} - \frac{4\sqrt{3}}{25}|S|^2\right). \tag{4.83}$$

Then the analyzing power for the aligned beam is given as

$$^TT_{20} \equiv -\frac{1}{2}(T_{20} + \sqrt{6}T_{22})$$

$$= -\frac{1}{N_R}\left(\text{Re}\{UT_{2\alpha}^* + \sqrt{3}UT_{2\beta}^*\} - \frac{16}{25}|S|^2\right). \tag{4.84}$$

The third-rank tensor analyzing powers iT_{31}, iT_{32} and $^TT_{30}$ are given as

$$iT_{31} = -\frac{4}{\sqrt{5}N_R}\text{Im}\{A^*B + C^*D + E^*F + G^*H - \sqrt{3}(B^*C + F^*G)\}$$

$$= \frac{4}{\sqrt{5}N_R}\text{Im}\left\{ UT_{3\alpha}^* \cot\theta + \frac{\sqrt{3}}{2}UT_{3\beta}^* - \frac{\sqrt{3}}{5}ST_{2\alpha}^* \right.$$

$$\left. - \frac{1}{10}ST_{2\beta}^* - \frac{5}{8}T_{2\alpha}T_{2\beta}^* \tan\theta \right\}, \tag{4.85}$$

$$iT_{32} = \frac{2\sqrt{2}}{N_R}\text{Im}\{A^*C + E^*G - F^*H - B^*D\}$$

$$= \frac{\sqrt{2}}{N_R}\text{Im}\left\{ -UT_{3\alpha}^* + \frac{1}{5}ST^* - T_{2\alpha}T_{2\beta}^* \right\} \tag{4.86}$$

and

$$^TT_{30} \equiv -\frac{1}{2}(\sqrt{3}iT_{31} + \sqrt{5}iT_{33})$$

$$= \frac{1}{\sqrt{5}N_R}\text{Im}\left\{ -2\sqrt{3}UT_{3\alpha}^* \cot\theta - 8UT_{3\beta}^* + \frac{6}{5}ST_{2\alpha}^* + \frac{6}{5}\sqrt{3}ST_{2\beta}^* \right\}. \tag{4.87}$$

Among these observables, we will investigate the second-rank tensor analyzing powers through Eqs. (4.81)–(4.83). In the case of deuteron scattering, $\sqrt{\frac{3}{2}}T_{20} - T_{22}$ and T_{21} give the measure of the T_R-type tensor interaction. As will be seen in Chapter 7, the T_R-type tensor interaction is more important in the ^{7}Li scattering. In the ^{7}Li case, these quantities are given as

$$\sqrt{\frac{3}{2}}T_{20} - T_{22} = \frac{1}{N_R}\left\{ \sqrt{2}\frac{\text{Re}(U^*T)}{\tan\theta} - \frac{4\sqrt{2}}{5}\text{Re}(S^*T) \right\} \tag{4.88}$$

and

$$2T_{21} = -\frac{1}{N_R}\left\{ \sqrt{2}\,\text{Re}(U^*T) + \frac{4\sqrt{2}}{5\tan\theta}\text{Re}(S^*T) \right\}, \tag{4.89}$$

both of which will be good measures of the tensor interaction, since they are proportional to the tensor amplitude T.

Further, $\text{Re}(U^*T)$ and $\text{Re}(S^*T)$ can be obtained by solving the above equations when the tensor analyzing powers T_{20}, T_{21} and T_{22}

are measured. At present, we will examine these equations by the use of the experimental data for ^{7}Li+^{58}Ni scattering at $E_{\text{Li}} = 14.22$ MeV [14]. For simplicity, we define T_{TU} and T_{ST} by

$$T_{TU} \equiv \frac{1}{2N_R}\text{Re}(U^*T) \tag{4.90}$$

and

$$T_{ST} \equiv -\frac{4}{5N_R}\text{Re}(S^*T). \tag{4.91}$$

Then Eq. (4.89) becomes

$$T_{21} = -\sqrt{2}T_{TU} + \frac{1}{\sqrt{2}}T_{ST}\cot\theta. \tag{4.92}$$

It is speculated that in the right-hand side of (4.92), $-\sqrt{2}T_{TU}$ is dominant because of the weakness of the vector interaction. In Fig. 4.2, we compare the measured T_{21} [14] with $-\sqrt{2}T_{TU}$ to examine the contribution of the T_{ST} term, where $-\sqrt{2}T_{TU}$ is evaluated from the experimental data of T_{20}, T_{21} and T_{22} [14] using the following formula obtained from Eqs. (4.88) and (4.89).

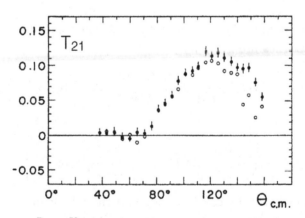

Fig. 4.2 T_{21} in ^{7}Li + ^{58}Ni elastic scattering at $E_{\text{Li}} = 14.22$ MeV. The closed circles indicate values of T_{21}. The open circles are $-\sqrt{2}T_{TU}$. The data are from Ref. [14].

$$-\sqrt{2}T_{TU} = -\frac{1}{2}\sin\theta\left\{\left(\sqrt{\frac{3}{2}}T_{20} - T_{22}\right)\cos\theta - 2T_{21}\sin\theta\right\}.$$

$$(4.93)$$

In the figure, $-\sqrt{2}T_{TU}$ reproduces well the measured T_{21} up to $\theta \approx 100°$, suggesting the weakness of the vector interaction in comparison with the central interaction. However, the discrepancy between $-\sqrt{2}T_{TU}$ and the measured T_{21} becomes systematically remarkable at $\theta \gtrsim 110°$ indicating the contribution of the T_{ST} term to be appreciable. This means that the vector interaction is weak but cannot be neglected at large angles.

Finally, we will present the transition amplitudes for the inelastic scattering to the excited state of ^{7}Li. The T-matrix M for the transition from the ground state ($\frac{3}{2}^-$) to the $\frac{1}{2}^-$ excited state is given by Ref. [30],

$$M = \begin{pmatrix} A & B & C & D \\ -D & C & -B & A \end{pmatrix},$$

$$(4.94)$$

where the row and the column are designated by the z-components of the spins, i.e. by $s_z = \frac{3}{2}, \frac{1}{2}, -\frac{1}{2}, -\frac{3}{2}$ from left to right for the ground state, and for the excited state, by $s_z = \frac{1}{2}$ and $-\frac{1}{2}$ from the top to the bottom. The matrix elements, $A \sim D$ can be described by the invariant amplitudes $F(s_i s_f K l_i, \cos\theta)$. For convenience, we will define the amplitudes of the vector S and the tensors $T_{2\alpha}$, $T_{2\beta}$ and $T_{2\gamma}$ as

$$S = F\left(\frac{3}{2}\frac{1}{2} 1 1; \cos\theta\right)\sin\theta, \qquad (4.95)$$

$$T_{2\alpha} = \left\{\sqrt{6}F\left(\frac{3}{2}\frac{1}{2} 2 0; \cos\theta\right)\cos\theta + F\left(\frac{3}{2}\frac{1}{2} 2 1; \cos\theta\right)\right\}\sin\theta,$$

$$(4.96)$$

$$T_{2\beta} = \sqrt{\frac{1}{8}} F \left(\frac{3}{2} \frac{1}{2} \ 2 \ 0; \cos \theta \right) (3 \cos^2 \theta - 1)$$

$$+ \sqrt{\frac{1}{3}} F \left(\frac{3}{2} \frac{1}{2} \ 2 \ 1; \cos \theta \right) \cos \theta + \sqrt{\frac{1}{2}} F \left(\frac{3}{2} \frac{1}{2} \ 2 \ 2; \cos \theta \right)$$

$$(4.97)$$

and

$$T_{2\gamma} = \sqrt{\frac{3}{8}} F \left(\frac{3}{2} \frac{1}{2} \ 2 \ 0; \cos \theta \right) \sin^2 \theta. \tag{4.98}$$

Using the above, we get the matrix elements of M in terms of S, $T_{2\alpha}$, $T_{2\beta}$ and $T_{2\gamma}$ as

$$A = \frac{1}{4} (\sqrt{3} S - T_{2\alpha}), \tag{4.99}$$

$$B = T_{2\beta}, \tag{4.100}$$

$$C = \frac{1}{4} (S + \sqrt{3} T_{2\alpha}) \tag{4.101}$$

and

$$D = T_{2\gamma}. \tag{4.102}$$

Using Eq. (3.18), one gets the scattering observable in terms of $A, \ldots, D$

$$\frac{d\sigma}{d\Omega} = \frac{1}{4} \frac{k_f}{k_i} N_R \tag{4.103}$$

with

$$N_R = 2(|A|^2 + |B|^2 + |C|^2 + |D|^2), \tag{4.104}$$

$$i T_{11} = \frac{2\sqrt{6}}{\sqrt{5}} \frac{1}{N_R} \text{Im} \left\{ AB^* + \frac{2}{\sqrt{3}} BC^* + CD^* \right\}, \tag{4.105}$$

$$T_{20} = \frac{2}{N_R} \{ |A|^2 - |B|^2 - |C|^2 + |D|^2 \}, \tag{4.106}$$

$$T_{21} = \frac{2\sqrt{2}}{N_R} \text{Re}\{-AB^* + CD^*\}, \tag{4.107}$$

$$T_{22} = \frac{2\sqrt{2}}{N_R} \text{Re}\{AC^* + BD^*\}, \tag{4.108}$$

$$iT_{31} = \frac{4}{\sqrt{5}N_R} \text{Im}\{AB^* - \sqrt{3}BC^* + CD^*\}, \tag{4.109}$$

$$iT_{32} = -\frac{2\sqrt{2}}{N_R} \text{Im}\{AC^* - BD^*\}, \tag{4.110}$$

$$iT_{33} = \frac{4}{N_R} \text{Im}\{AD^*\}. \tag{4.111}$$

The representations in spherical amplitudes are obtained by the use of Eqs. (4.99)–(4.102), which will be discussed in Chap. 7.

Chapter 5

Optical Potential and Elastic Scattering of Protons

In the following three chapters, we will investigate the scattering of protons, deuterons and ^{7}Li by nuclei, through numerical calculations, where the interactions between the projectile and the target are derived based on folding models. In the cases of deuteron and ^{7}Li, the contributions of the virtual excitations of the projectile are taken into account by the coupled-channel method.

5.1 Folding Model Interaction between Proton and Nucleus

The proton–nucleus interaction will be obtained by folding nuclear forces between the projectile proton and a nucleon of the nucleus, over the nucleus. The main part of the force will be produced by the one-pion exchange mechanism. Assign suffixes o and i to the projectile proton and the target nucleon and denote their relative coordinate by r_{oi}. The pion exchange force will then be given by

$$v_{oi} = v_d(r_{oi}) + v_\tau(r_{oi})(\boldsymbol{\tau}_o \cdot \boldsymbol{\tau}_i) + v_\sigma(r_{oi})(\boldsymbol{\sigma}_o \cdot \boldsymbol{\sigma}_i)$$
$$+ v_{\sigma\tau}(r_{oi})(\boldsymbol{\sigma}_o \cdot \boldsymbol{\sigma}_i)(\boldsymbol{\tau}_o \cdot \boldsymbol{\tau}_i) + \{v_t(r_{oi}) + v_{t\tau}(r_{oi})(\boldsymbol{\tau}_o \cdot \boldsymbol{\tau}_i)\}s_{oi},$$
$$(5.1)$$

where s_{oi} is the tensor operator,

$$s_{oi} = 3\frac{(\boldsymbol{r}_{oi} \cdot \boldsymbol{\sigma}_o)(\boldsymbol{r}_{oi} \cdot \boldsymbol{\sigma}_i)}{r_{oi}^2} - (\boldsymbol{\sigma}_o \cdot \boldsymbol{\sigma}_i). \qquad (5.2)$$

49

For a spherical target, the terms proportional to $\boldsymbol{\sigma}_i$ are neglected because they will vanish by the folding procedure, as $\sum_i \boldsymbol{\sigma}_i = 0$. Further, we will replace the nucleon in the nucleus by the nuclear matter, the density of which is denoted by ρ_p and ρ_n for the proton and the neutron, respectively. We also set the origin of the coordinate axes to be at the center of the nucleus. Then the folding potential U is given by

$$U = U_R + U_S, \tag{5.3}$$

where

$$U_R = \int v_d(|\boldsymbol{r} - \boldsymbol{r}'|)\rho_m(r')dr' \quad \text{with} \quad \rho_m = \rho_p + \rho_n \tag{5.4}$$

and

$$U_S = \tau_z \int v_\tau(|\boldsymbol{r} - \boldsymbol{r}'|)\{\rho_p(r') - \rho_n(r')\}dr'. \tag{5.5}$$

Further approximations are applied to the nuclear force

$$v_\tau(r) = -\zeta v_d(r) \tag{5.6}$$

and to the nucleon density

$$\rho_p = \frac{Z}{A}\rho_m \quad \text{and} \quad \rho_n = \frac{N}{A}\rho_m. \tag{5.7}$$

Then we get

$$U = \left(1 + \zeta\frac{N - Z}{A}\tau_z\right) U_R. \tag{5.8}$$

In addition to the above, some corrections will be considered. First, it has been suggested that exchanges of heavy mesons between nucleons produce spin-orbit interactions, which will be included as a phenomenological potential between the proton and the nucleus. Second, the many-body nature of the target nucleus may open reaction channels by collision with the incident proton. In the elastic-scattering channel, this effect will be counted as absorption of the beam and is taken into account by the introduction of imaginary parts of the interactions. Third, the Coulomb interactions between the incident proton and the protons of the target nucleus are treated by regarding the target nucleus as a uniformly charged sphere. The interaction which includes the above corrections is called the optical model potential [15] because of the existence of the imaginary parts, in analogy with optics.

5.2 Scattering of Proton by Optical Model Potential

In elastic scattering of nucleons by nuclei, the optical model potential discussed above has been used as the standard prescription for analyses of experimental data. Practically, the potential is given as the function of the nucleon–nucleus relative coordinate r with the following form factor,

$$U(r) = V_c - V_o f(x_o) + \left(\frac{1}{m_\pi}\right)^2 V_{so}(\boldsymbol{\sigma} \cdot \boldsymbol{l}) \frac{1}{r} \frac{d}{dr} f(x_{so})$$
$$- i \left\{ W f(x_W) - 4 W_D \frac{d}{dx_D} f(x_D) \right\}, \tag{5.9}$$

where $\boldsymbol{\sigma}$ is the Pauli spin operator of the nucleon, $\boldsymbol{l}$ is the nucleon–nucleus relative orbital angular momentum and the parameter $(\frac{1}{m_\pi})^2$ is fixed to be 2, as usual. In the zero-range limit of v_d, $U(r)$ has a functional form similar to the nuclear matter density ρ_m, which is well reproduced by the so-called Woods–Saxon type function. Then $f(x_i)$ is given by

$$f(x_i) = \frac{1}{1 + e^{x_i}} \tag{5.10}$$

with

$$x_i = \frac{r - r_i A^{\frac{1}{3}}}{a_i}, \tag{5.11}$$

where $i = o$, so, w and w_D. The quantities V_o and V_{so} are the depths of the central and spin-orbit interactions and W and W_D are the depths of the imaginary parts, for the body and the surface types. At low incident energies, the absorption of the beam will occur at the nuclear surface, while at high energies body-type absorption will become important. Sometimes, absorption effects are considered also for the spin-orbit interaction. The term V_c describes the Coulomb potential,

$$V_c = \frac{zz'e^2}{r}, \quad r \geq R_c \tag{5.12}$$

and

$$V_c = \frac{zz'e^2}{2R_c}\left(3 - \frac{r^2}{R_c^2}\right), \quad r \leq R_c \tag{5.13}$$

with

$$R_c = r_c A^{\frac{1}{3}}, \tag{5.14}$$

where z is the atomic number of the target, $z' = 1$ for the proton and $z' = 0$ for the neutron. $r_i A^{\frac{1}{3}}$ describes the nuclear radius and a_i describes the diffuseness of the target nucleus.

The depth, radius and diffuseness of each potential are treated as parameters which should be determined by fitting the experimental data of the differential cross section and the vector analyzing power. Such parameter searches have been performed at many incident energies and give the parameters as functions of the incident energy E. Here, we will show two typical sets of the parameters for the proton; the first, by Becchetti and Greenlees [16], which is one of the most popular sets, and the other by Menet [17], which has been refined by including the reaction cross section data.

(i)   BG potential ($A > 40$ and $E \lesssim 50\,\text{MeV}$)

$$V_0 = 54.0 - 0.32E + 24\frac{(N-Z)}{A} + 0.4\frac{Z}{A^{\frac{1}{3}}}, \tag{5.15}$$

$$r_0 = 1.17, \quad a_0 = 0.75, \tag{5.16}$$

$$W = 0.22E - 2.7, \quad (\text{set to zero, if negative}) \tag{5.17}$$

$$W_D = 11.8 - 0.25E + 12\frac{N-Z}{A}, \quad (\text{set to zero, if negative}) \tag{5.18}$$

$$r_W = r_D = 1.32, \tag{5.19}$$

$$a_W = a_D = 0.51 + 0.7\frac{(N-Z)}{A}, \tag{5.20}$$

$$V_{so} = 6.2, \quad r_{so} = 1.01, \quad a_{so} = 0.75. \tag{5.21}$$

(ii) Menet potential ($E = 30 \sim 60 \, \text{MeV}$)

$$V_0 = 49.9 - 0.22E + 26.4\frac{N - Z}{A} + 0.4\frac{Z}{A^{\frac{1}{3}}}, \tag{5.22}$$

$$r_0 = 1.16, \quad a_0 = 0.75, \tag{5.23}$$

$$W = 1.2 + 0.09E, \tag{5.24}$$

$$W_D = 4.2 - 0.05E + 15.5\frac{N - Z}{A}, \tag{5.25}$$

$$r_W = r_D = 1.37, \tag{5.26}$$

$$a_W = a_D = 0.74 - 0.008E + 1.0\frac{N - Z}{A}, \tag{5.27}$$

$$V_{so} = 6.04, \quad r_{so} = 1.064, \quad a_{so} = 0.78, \tag{5.28}$$

$$r_c = 1.25, \tag{5.29}$$

where the potential depth and the incident energy are expressed in MeV and the length in fm.

The above potential parameters become averaged over various target nuclei. For the individual target, to determine the potential parameters in an unambiguous way, we have to take into account different kinds of observables in the analysis. In this sense, one should include, in the analysis, the Wolfenstein parameters [18] in addition to the cross section and the vector analyzing power. Let us consider a new coordinate system (x', y', z') which is obtained by rotating the (x, y, z) system by angle θ around the y-axis. Referring to these systems, the Wolfenstein parameters R and Q are defined as

$$R = K_x^{x'}, \tag{5.30}$$

$$Q = \sqrt{1 - P_y'^2}\sin\beta \quad \text{with} \quad \beta = \tan^{-1}\frac{K_z^x}{K_x^x}. \tag{5.31}$$

The measurements, which include R and Q, have been performed at $E_p = 65$ MeV. In Fig. 5.1, results are compared with the calculation by the parameter sets of Table 5.1. Clearly, good agreement is obtained between the results of experiment and calculation.

Table 5.1 Potential Parameters for ^{40}Ca and ^{90}Zr.

For ^{40}Ca target	For ^{90}Zr target
$V_0 = 39.922$, $r_0 = 1.102$, $a_0 = 0.744$,	$V_0 = 33.617$, $r_0 = 1.251$, $a_0 = 0.704$,
$W = 9.417$, $r_W = 0.970$, $a_W = 0.252$,	$W = 9.515$, $r_W = 0.898$, $a_W = 0.358$,
$W_D = 6.452$, $r_D = 1.214$, $a_D = 0.588$,	$W_D = 6.435$, $r_D = 1.198$, $a_D = 0.620$,
$V_{so} = 5.550$, $r_{so} = 0.950$, $a_{so} = 0.603$,	$V_{so} = 4.423$, $r_{so} = 1.078$, $a_{so} = 0.615$,
$W_{so} = -0.192$, $r_{wso} = 1.769$.	$W_{so} = -0.230$, $r_{wso} = 1.492$.

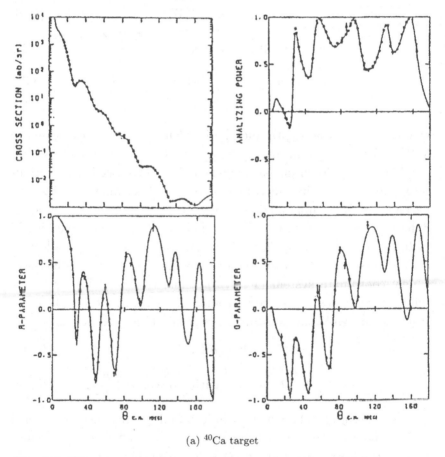

(a) ^{40}Ca target

Fig. 5.1 The comparison between the measured and the calculated cross section, analyzing power, R-parameter and Q-parameter for p-nucleus elastic scattering at $E_p = 65$ MeV. The target nucleus is ^{40}Ca in (a) and is ^{90}Zr in (b).

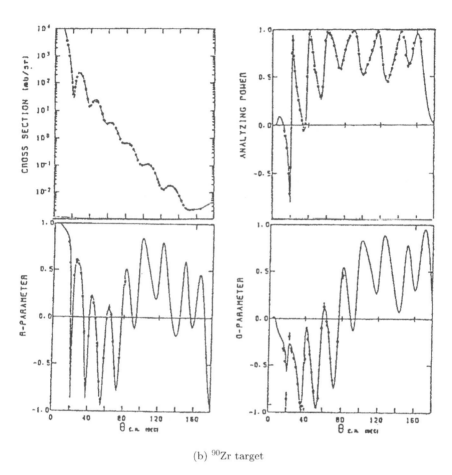

(b) ^{90}Zr target

Fig. 5.1 (*Continued*)

Chapter 6

Folding Model Interaction and Virtual Excitation in Scattering of Deuterons

In Sec. 4.2, we have studied the interaction between a spin-1 particle and a spin-less nucleus and found that it consists of central, spin-orbit and second-rank tensor interactions. Here, we will derive these interactions for deuterons in the folding model, through which the scattering by nuclei is studied. Further corrections which arise from virtual excitations of the deuteron internal motion are investigated. For numerical calculations, the Continuum-Discretized Coupled Channel (CDCC) method [20] is employed.

6.1 Folding Model for Deuteron Nucleus Interaction

As was shown in the preceding chapter, the optical model potential for the interaction between a nucleon and a nucleus produces successful agreement with experimental data in proton–nucleus elastic scattering. Since the deuteron is a loosely-bound system of a proton and a neutron, in order to obtain the deuteron–nucleus optical potential, it is reasonable to fold such proton–nucleus and neutron–nucleus optical potentials into the deuteron ground state.

Denote the deuteron internal wave function by ψ_d, which consists of the S-state (ψ_S) and the D-state (ψ_D), so that

$$\psi_d = \psi_S + a_D \psi_D \qquad (6.1)$$

57

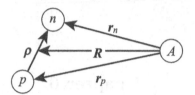

Fig. 6.1   Coordinates of deuteron.

with

$$\psi_S = \chi_{1v} Y_{00}(\hat{\rho}) u_S(\rho) \tag{6.2}$$

and

$$\psi_D = \sum_{v'} (12v'm \mid 1v)\chi_{1v'} Y_{2m}(\hat{\rho}) u_D(\rho), \tag{6.3}$$

where ρ is the proton–neutron relative coordinate as shown in Fig. 6.1 and $\hat{\rho}$ is the angular part of ρ. Here χ_{1v} is the wave function of the intrinsic spin with the spin z component v. The coefficient a_D describes the amplitude of the D-state. Since $|a_D|^2$ is very small as 0.04–0.07, we will neglect the $|a_D|^2$ term in the folding procedure.

First, we will calculate the two contributions to the folding potential: the central part of the proton–nucleus (A) optical potential, $U_{PA}^{(c)}$, and that of the neutron–nucleus optical potential, $U_{nA}^{(c)}$. The potential thus obtained consists of two components, $U_{dA}(S^2)$ which is obtained by the S-state of the deuteron ground state and $U_{dA}(SD)$ which is obtained as the interference term of the S and D states,

$$U_{dA}(S^2) = \int \psi_S^* \left\{ U_{pA}^{(c)} \left(\mathbf{R} - \frac{1}{2}\boldsymbol{\rho} \right) + U_{nA}^{(c)} \left(\mathbf{R} + \frac{1}{2}\boldsymbol{\rho} \right) \right\} \psi_S d\boldsymbol{\rho} \tag{6.4}$$

and

$$U_{dA}(SD) = a_D \left\{ \int \psi_S^* \left[U_{pA}^{(c)} \left(\mathbf{R} - \frac{1}{2}\boldsymbol{\rho} \right) + U_{nA}^{(c)} \left(\mathbf{R} + \frac{1}{2}\boldsymbol{\rho} \right) \right] \psi_D d\boldsymbol{\rho} \right.$$

$$\left. + (\psi_S \Leftrightarrow \psi_D) \text{ term} \right\}, \tag{6.5}$$

where $\boldsymbol{R}$ is the deuteron-A relative coordinate as shown in Fig. 6.1. As will be discussed in the following, $U_{dA}(S^2)$ is the central interaction and $U_{dA}(SD)$ is the second-rank tensor interaction.

In Eq. (6.4), to perform the integration on $\boldsymbol{\rho}$, we will expand $U_{pA}^{(c)}$ and $U_{nA}^{(c)}$ in the multipoles, for example for $U_{nA}^{(c)}$:

$$U_{nA}^{(c)}\left(\boldsymbol{R}+\frac{1}{2}\boldsymbol{\rho}\right) = 4\pi \sum_\lambda \sum_m \frac{(-)^m}{2\lambda+1} V_{nA}^{(\lambda)}(R,\rho) Y_{\lambda m}(\hat{R}) Y_{\lambda-m}(\hat{\rho}).$$

(6.6)

Using Eq. (6.6), we get

$$\int \psi_S^*[U_{pA}^{(c)} + U_{nA}^{(c)}]\psi_S d\boldsymbol{\rho} = \langle u_S|V_{pA}^{(0)}(R,\rho) + V_{nA}^{(0)}(R,\rho)|u_S\rangle_\rho,$$

(6.7)

where $\langle\ \rangle_\rho$ means the integral over ρ. Equation (6.7) shows that $U_{dA}(S^2)$ gives the d-A central interaction. Further, $U_{dA}(SD)$ is calculated as

$$a_D \int \psi_S^*[U_{pA}^{(c)} + U_{nA}^{(c)}]\psi_D d\boldsymbol{\rho}$$

$$= \frac{4\pi}{5} a_D \langle u_S|V_{pA}^{(2)}(R,\rho) + V_{nA}^{(2)}(R,\rho)|u_D\rangle_\rho (12\nu0 \mid 1\nu) Y_{20}(\hat{R}).$$

(6.8)

On the other hand, one obtains the matrix element of the T_R-type tensor interaction between the S-states with the spin-direction ν. By utilizing $[\boldsymbol{R}\otimes\boldsymbol{R}]_\mu^2 = R^2 Y_{2\mu}(\hat{R})$ and applying the Wigner–Eckart theorem [3] for $\langle 1\nu|[\boldsymbol{s}\otimes\boldsymbol{s}]_0^2|1\nu\rangle$, we get

$$\langle \psi_S|T_R \mid \psi_S\rangle = \frac{1}{\sqrt{3}}(1\|[\boldsymbol{s}\otimes\boldsymbol{s}]^2\|1)R^2 V(R)(12\nu0 \mid 1\nu)Y_{20}(\hat{R}). \quad (6.9)$$

When the shape and magnitude of $V(R)$ are properly controlled, Eq. (6.9) will agree with Eq. (6.8), showing that the D-state contribution produces the T_R-type tensor interaction.

Next, we will investigate folding interactions due to the spin-orbit (LS) parts of the p-A and n-A potentials. Before the folding procedure, the orbital angular momenta of the proton and the neutron relative to the target A, l_p and l_n respectively, are transformed to the angular momentum of the d-A relative motion L and the deuteron internal angular momentum l in turn. Transforming the coordinate system from r_p and r_n to R and ρ as in Fig. 6.1,

$$r_p = R - \frac{1}{2}\rho \quad \text{and} \quad r_n = R + \frac{1}{2}\rho. \tag{6.10}$$

Correspondingly,

$$\nabla_p = \frac{1}{2}\nabla_R - \nabla_\rho \quad \text{and} \quad \nabla_n = \frac{1}{2}\nabla_R + \nabla_\rho. \tag{6.11}$$

Using the above, one calculates the orbital angular momenta, l_p and l_n, by the equations

$$l_p = \frac{1}{i}r_p \times \nabla_p \quad \text{and} \quad l_n = \frac{1}{i}r_n \times \nabla_n. \tag{6.12}$$

The proton LS potential $U_{pA}^{(so)}(r_p)l_p \cdot \sigma_p$ and the neutron LS potential $U_{nA}^{(so)}(r_n)l_n \cdot \sigma_n$ are transformed by Eqs. (6.10)–(6.12) to deuteron–nucleus interactions. Defining L and l as

$$L = \frac{1}{i}R \times \nabla_R \quad \text{and} \quad l = \frac{1}{i}\rho \times \nabla_\rho, \tag{6.13}$$

we get

$$U_{pA}^{(so)}(r_p)l_p \cdot \sigma_p + U_{nA}^{(so)}(r_n)l_n \cdot \sigma_n$$

$$= \frac{1}{2}U_+^{(so)}(L+l) \cdot s + \frac{1}{i}U_-^{(so)}\left(R \times \nabla_\rho + \frac{1}{4}\rho \times \nabla_R\right) \cdot s$$

$$+ \frac{1}{2}U_-^{(so)}(L+l) \cdot s' + \frac{1}{i}U_+^{(so)}\left(R \times \nabla_\rho + \frac{1}{4}\rho \times \nabla_R\right) \cdot s' \tag{6.14}$$

with

$$U_\pm^{(so)} = U_{pA}^{(so)}(r_p) \pm U_{nA}^{(so)}(r_n) \tag{6.15}$$

and

$$s = \frac{1}{2}(\sigma_p + \sigma_n), \quad s' = \frac{1}{2}(\sigma_p - \sigma_n), \tag{6.16}$$

where $U_{pA}^{(so)}$ and $U_{nA}^{(so)}$ are the form factors of the LS part of the p-A potential and that of the n-A potential, respectively. In the folding, the dominant contribution arises from the S-state of the deuteron internal motions, and one can then neglect the terms linear to l, ρ and ∇_ρ in Eq. (6.14). Further, the s' term does not contribute to the folding potential because the operator s' changes the spin of the p-n system, from 1 to 0 or from 0 to 1 in the first order. The second-order contribution will be discussed later. Finally, we get the spin-orbit potential of the deuteron as

$$
U_{dA}(\text{LS}) = \frac{1}{2} \left\{ \langle u_S | U_{pA}^{(so)} \left(\left| \boldsymbol{R} - \frac{1}{2} \boldsymbol{\rho} \right| \right) | u_S \rangle_\rho \right.
$$

$$
\left. + \langle u_S | U_{nA}^{(so)} \left(\left| \boldsymbol{R} + \frac{1}{2} \boldsymbol{\rho} \right| \right) | u_S \rangle_\rho \right\} (\boldsymbol{L} \cdot \boldsymbol{S}). \qquad (6.17)
$$

6.2 Interaction Induced by Virtual Excitation

During the scattering by nuclei, the deuteron projectile may be excited to continuum states by interaction with the target, followed by de-excitation to the ground state. Such virtual excitation can be treated quantitatively by the CDCC method which will be explained in detail in the next section. At present, we will qualitatively investigate the nature of effective interactions induced by such virtual excitations.

The effective interaction $V^{\textit{eff}}$ due to the virtual excitation is described typically in the second-order perturbation approximation. The interaction $V^{\textit{eff}}$ by the central part of the nucleon-target potential $V(c)(\boldsymbol{R}, \boldsymbol{\rho})$, which is expected to give the largest contribution to the effective interaction, is given as

$$
V^{\textit{eff}} = \sum_k \frac{\langle u_S | V(c)(\boldsymbol{R}, \boldsymbol{\rho}) | u_{\boldsymbol{k}} \rangle \langle u_{\boldsymbol{k}} | V(c)(\boldsymbol{R}, \boldsymbol{\rho}) | u_S \rangle}{-\epsilon_d - \frac{\hbar^2}{m} k^2}, \qquad (6.18)
$$

where $u_{\boldsymbol{k}}$ represents the continuum wave function of the p-n relative momentum $\boldsymbol{k}$. In the ground state, the D-state admixture which has small contributions is neglected, for simplicity. The interaction $V(c)$

is given by

$$V(c)(\boldsymbol{R}, \boldsymbol{\rho}) = V_{pA}^{(c)}\left(\left|\boldsymbol{R} - \frac{1}{2}\boldsymbol{\rho}\right|\right) + V_{nA}^{(c)}\left(\left|\boldsymbol{R} + \frac{1}{2}\boldsymbol{\rho}\right|\right). \tag{6.19}$$

To see the global nature of V^{eff}, we will approximate the Green's function

$$-\frac{1}{\epsilon_d + \hbar^2 k^2/m}$$

by that for a particular $\boldsymbol{k}$, for instance $\boldsymbol{k} = 0$. Hereafter, we call this the adiabatic approximation. Using

$$\sum_{\boldsymbol{k}} |u_{\boldsymbol{k}}\rangle\langle u_{\boldsymbol{k}}| = 1, \tag{6.20}$$

we get for the adiabatic approximation

$$V^{eff} = -\frac{1}{\epsilon_d}\langle u_S|(V(c)(\boldsymbol{R}, \boldsymbol{\rho}))^2|u_S\rangle_\rho, \tag{6.21}$$

which will give a central interaction for the deuteron center-of-mass motion.

Next, we will consider the contribution of the spin-orbit interaction $V(LS)(\boldsymbol{R}, \boldsymbol{\rho})\boldsymbol{L} \cdot \boldsymbol{S}$, where

$$V(LS)(\boldsymbol{R}, \boldsymbol{\rho}) = \left\{U_{pA}^{(so)}\left(\left|\boldsymbol{R} - \frac{1}{2}\boldsymbol{\rho}\right|\right) + U_{nA}^{(so)}\left(\left|\boldsymbol{R} + \frac{1}{2}\boldsymbol{\rho}\right|\right)\right\}, \tag{6.22}$$

The effective interaction due to the virtual excitation by this interaction will be obtained in a way similar to Eqs. (6.18)–(6.21), so that

$$V^{eff} = -(1/\epsilon_d)\langle u_S|(V(LS)(\boldsymbol{R}, \boldsymbol{\rho}))^2|u_S\rangle(\boldsymbol{L} \cdot \boldsymbol{S})^2. \tag{6.23}$$

Further, $(\boldsymbol{L} \cdot \boldsymbol{S})^2$ is expanded in the spin-space tensors as

$$(\boldsymbol{L} \cdot \boldsymbol{S})^2 = \frac{2}{3}L^2 - \frac{1}{2}(\boldsymbol{L} \cdot \boldsymbol{S}) + ([\boldsymbol{S} \otimes \boldsymbol{S}]^2 \cdot [\boldsymbol{L} \otimes \boldsymbol{L}]^2). \tag{6.24}$$

The last term in the right-hand side of the above equation is the second-rank tensor operator of the T_L-type. That means, the virtual excitation of the continuum states by the LS interaction produces a T_L-type tensor interaction. This suggests that the virtual excitation by the $(\boldsymbol{L} \cdot \boldsymbol{S'})$ interaction, which appears in Eq. (6.14), also produces a second-rank tensor interaction of T_L-type, since

$$(\boldsymbol{L} \cdot \boldsymbol{S'})^2 = \frac{1}{3} L^2 - \frac{1}{2} (\boldsymbol{L} \cdot \boldsymbol{S}) - ([\boldsymbol{S} \otimes \boldsymbol{S}]^2 \cdot [\boldsymbol{L} \otimes \boldsymbol{L}]^2). \qquad (6.25)$$

However, the contribution of the $(\boldsymbol{L} \cdot \boldsymbol{S'})$ term has different characteristics from that of the $(\boldsymbol{L} \cdot \boldsymbol{S})$ term. The intermediate states effective for the $(\boldsymbol{L} \cdot \boldsymbol{S})$ interaction are spin-triplet states, for example, of S and D states, while those effective for the $(\boldsymbol{L} \cdot \boldsymbol{S'})$ interaction are the spin-singlet states, for example, of P states. Thus, in quantitative calculations, these contributions are different from each other and will be easily identified.

6.3 Quantitative Analysis by the CDCC Method

As mentioned in the previous section, we will treat the scattering of deuterons by the CDCC method, taking into account the contribution of the virtual excitation to the continuum. In the following, we will give the theoretical development of the method and then discuss the result of the numerical calculation.

For simplicity, we will tentatively consider two channels, the elastic one and the inelastic one where the projectile stays in an excited state. When the interaction is strong, these two channels interact with each other and have to be solved simultaneously. This is performed by a coupled-channel method.

When the projectile is a deuteron, the excited states form a continuum spectrum of positive energy. To include these states in the coupled-channel scheme, we will discretize the continuum by dividing into a finite number (N) of momentum bins of $\Delta \boldsymbol{k}$.

The ith momentum bin is designated by the representative momentum $\boldsymbol{k}_i$ and is included as a component in the coupled-channel scheme. The magnitude of N is restricted by the upper limit of $\boldsymbol{k}$, $k_{\max}$, as $N = k_{\max}/\Delta k$, and $k_{\max}$ and Δk are determined so that the results of the coupled-channel calculation converge. Such a procedure is called the Continuum Discretized Coupled Channel (CDCC)

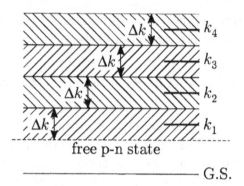

Fig. 6.2   Discretization of the continuum state for deuteron. G.S. refers to the ground state of the deuteron.

method [20]. In the following, we will apply the method to elastic scattering of deuterons by nuclei.

Let us assume the spin of the nucleus to be zero. Then the wave function of the total system, Ψ_{JM}, which has the total angular momentum J and z-component M, is given by

$$\Psi_{JM} = \sum_{L=|J-I|}^{J+1} [\Phi_0(\boldsymbol{\rho}) \otimes \phi_0(L, J, : \boldsymbol{R})]_{JM} + \sum_{l=0}^{\infty} \sum_{S=0}^{1}$$

$$\times \sum_{I=|l-S|}^{l+S} \sum_{L=|J-I|}^{J+I} \sum_{i=1}^{N} [\Phi(^{2S+1}l_I; k; \boldsymbol{\rho}) \otimes \phi(^{2S+1}l_I, L, J; \boldsymbol{R})]_{JM}.$$

$$(6.26)$$

In the above, the first term describes the elastic-scattering channel where the deuteron is in the ground state, while in the second term the deuteron is excited to the continuum. The deuteron intrinsic spin is denoted by s and $2s + 1$ (=1 or 3) distinguishs the singlet state from the triplet one. The coordinate $\boldsymbol{\rho}$ and the associated angular momentum l are for the deuteron internal motion, and $I = s+l$ gives the total angular momentum of the deuteron. The deuteron–target relative coordinate and orbital angular momentum are denoted by $\boldsymbol{R}$ and $\boldsymbol{L}$, respectively. The $\boldsymbol{J} = \boldsymbol{L} + \boldsymbol{I}$ stands for the total angular momentum of the system. Inserting Eq. (6.26) into the

Schrödinger equation,

$$(H - E)\Psi_{JM} = 0, \tag{6.27}$$

we get the simultaneous equation for the ground state and the excited states. After the integration on the angular variables, the equation for the radial wave function u is derived to be

$$\left\{ -\frac{1}{2\mu_d}\frac{d^2}{dR^2} + \frac{L(L+1)}{2\mu_d R^2} + V_{\mathrm{Coul}}(R) - E_c \right\} u_{cL}^J(R)$$

$$= -\sum_{c'L'} F_{cL;c'L'}^{(J)}(R)u_{c'L'}^J(R), \tag{6.28}$$

where c's designate channels; that is, c_0 stands for the deuteron ground state and c_1, c_2 etc. for the excited states. $F_{cL;c'L'}^{(J)}$, the interaction between the cL channel and $c'L'$ channel, is given by

$$F_{cL;c'L'}^{(J)}(R) = \langle i^L[\Phi_c(\boldsymbol{\rho}) \otimes Y_L(\hat{R})]_{JM}|U_p + U_n|i^{L'}[\Phi_{c'}(\boldsymbol{\rho}) \otimes Y_{L'}(\hat{R})]_{JM}\rangle, \tag{6.29}$$

where U_p and U_n are respectively the proton-target interaction and the neutron-target interaction. Solving the coupled-channel equation (6.28) under the boundary condition

$$u_{c'L'}^J(R)(R \to \infty) = \delta_{c'c_0}\delta_{L'L}U_{c'L'}^{(-)}(R) - \sqrt{\frac{v_0}{v'}}S_{c'L';c_0L}^J U_{c'L'}^{(+)}(R), \tag{6.30}$$

we get the S matrix $S_{c'L';c_0L}^J$ for the transition from the c_0L channel to the $c'L'$ channel. The functions $U_{c'L'}^{(+)}$, and $U_{c'L'}^{(-)}$ are the Coulomb functions in the asymptotic form of the outgoing wave and of the incoming wave, respectively.

For the elastic scattering, physical observables are calculated by the diagonal elements of S-matrix, $S_{c_0L;c_0L}^J$. In Fig. 6.3, the calculated observables are compared with the measured ones. There observables are the Rutherford ratio of the cross section σ/σ_R, the vector analyzing power A_y and the tensor analyzing powers A_{xx}, A_{yy}, A_{xz} for the ^{40}Ca target at $E_d = 56$ MeV. In the calculation, $k_{\mathrm{max}} = 1.0\,\mathrm{fm}^{-1}$ and $\Delta k = 0.25\,\mathrm{fm}^{-1}$ are employed for the discretization. For U_p

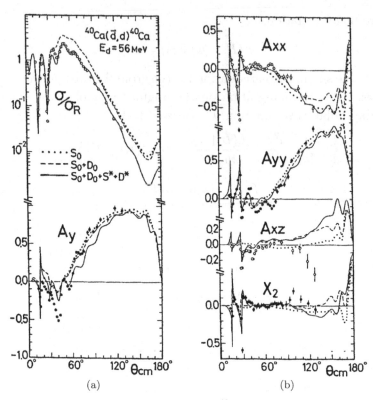

Fig. 6.3   Elastic scattering of deuteron from ^{40}Ca at $E_d = 56$ MeV. The CDCC calculation is compared with experimental data. S_0 and D_0 are contributions of the ground state and S^* and D^* are those of the excited states.

and U_n the Becchetti–Greenlees optical potential [16] is adopted, and for the p-n system the S and D waves are considered, which are calculated by the use of the Reid-soft-core potential [21]. In the figure, $X_2 = -\sqrt{\frac{1}{3}}(2A_{xx} + A_{yy})$ is also displayed. As discussed in Sec. 4.3, this quantity is a measure of the D-state contribution as well as A_{xz}.

The calculation in Ref. [22] is overall in agreement with the experimental data for the observables investigated. In the following, we will discuss the details of the calculation, focusing attention on the effects of the D-state in the ground state and those of the virtual excitation of the continuum states, S^* and D^*, by the channel coupling.

As investigated in the previous section, the virtual excitation of the S^* and D^* states by the central parts of the p-A and n-A optical potentials produces additional d-A central interactions. This provides large contributions to the cross section as seen in Fig. 6.3, and improves the agreement with experimental data. The sharp structures of the angular distribution of the analyzing powers at forward angles are mostly due to the reflection of the diffraction pattern of the cross section in their denominators. In the tensor analyzing powers, A_{xx} and A_{yy}, at a large angle, the contribution of the D-state is rather minor and the main part of the analyzing powers will be explained as the contribution of the T_L-type tensor interaction, as pointed out in Sec. 4.3. Such T_L-type tensor interactions are induced by multiple scattering of the deuteron due to the LS interaction. This is implicitly included in Eq. (6.28), where the second order of the LS interaction produces a T_L-type tensor interaction of the deuteron as discussed in the previous section. The measured A_{xz} and X_2 at $\theta \lesssim 90°$ are reproduced by including the D-state admixture in the deuteron ground state. The contribution of the D-state becomes large at larger angles but does not explain the experimental data.

To investigate the contribution of the virtual excitation of spin-singlet states in comparison with that of spin-triplet states, the CDCC calculations are performed for deuteron elastic scattering at $E_d = 80$ MeV and 400 MeV, which include the $^3S_1$, ^3D_J ($J = 1, 2, 3$) and $^1P_1$ break-up states for the deuteron as the spin-triplet and spin-singlet examples [23]. For the nucleon-target optical potentials, the standard Woods–Saxon potential is assumed for $E_d = 80$ MeV while the effective Schrödinger form of the Dirac phenomenology [24] is adopted for $E_d = 400$ MeV. The results of the calculation for the ^{58}Ni target are shown in Figs. 6.4 and 6.5. The calculated σ/σ_R, A_y, A_{yy} and X_2, which include the virtual excitations, reproduce the experimental data well except for α/σ_R at large angles. The contribution of the excitation of the spin-singlet state has an appreciable magnitude in A_{yy} and improves the agreement with the data at relatively large angles.

Finally, in Figs. 6.6(a) and 6.6(b), we present the magnitudes of the amplitudes U, S, T_α and T_β as functions of the scattering angle θ

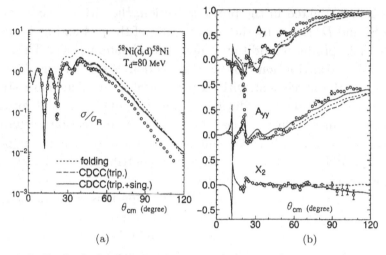

(a)   (b)

Fig. 6.4   Rutherford ratio of cross section σ/σ_R, vector analyzing power A_y and tensor analyzing powers A_{yy} and X_2 of elastic scattering $^{58}\text{Ni}(\vec{d}, d)^{58}\text{Ni}$ at $E_d = 80$ MeV. The dotted, dashed and solid lines are calculated by a simple folding model and by the CDCC method, which includes excitations to triplet states and to states which include both triplet and singlet.

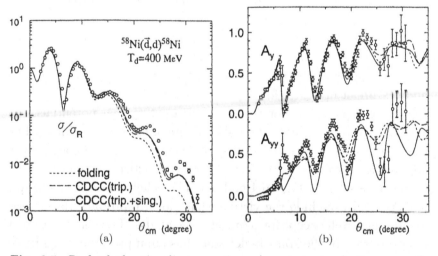

(a)   (b)

Fig. 6.5   Rutherford ratio of cross section σ/σ_R, vector analyzing power A_y and tensor analyzing powers A_{yy} and X_2 of elastic scattering $^{58}\text{Ni}(\vec{d}, d)^{58}\text{Ni}$ at $E_d = 400$ MeV. See the caption of Fig. 6.4, which indicates the definitions of the dotted, dashed and solid lines.

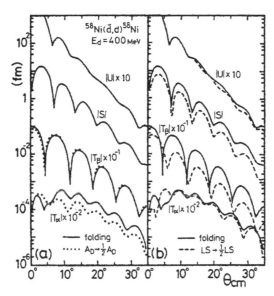

Fig. 6.6 Dependence of scattering amplitudes on a_D, the D-state admixture of the deuteron, and the strength of the LS interaction for the ^{58}Ni($\vec{d}$, d)^{58}Ni scattering at $E_d = 400$ MeV. $|U|$, $|S|$, $|T_\alpha|$ and $|T_\beta|$ are the absolute magnitudes of the scalar, vector, T_R-type tensor and T_L-type tensor amplitudes.

for the ^{58}Ni target at $E_d = 400$ MeV [23]. For simplicity, the virtual excitations of the continuum states are discarded. In the figure, we will examine the dependence of these amplitudes on the strength of the T_R tensor interaction as well as on the strength of the LS interaction. In (a), the calculation where the D-state amplitude of the deuteron is reduced by the factor $\frac{1}{2}$ is compared with the original calculation, and in (b), the calculation where the strength of the LS interaction is reduced by the factor $\frac{1}{2}$ is compared with that of the original.

As can be seen in the figure, in (a), $|T_\alpha|$ is reduced to about $\frac{1}{2}$ of the original, except for small angles, while other amplitudes remain unchanged by the reduction of the D-state, and in (b), $|S|$ is reduced by the factor $\frac{1}{2}$ and $|T_\beta|$ is reduced by the factor $\frac{1}{4}$, while for the reduction of the LS interaction $|T_\alpha|$ is almost unchanged except for the forward angles. $|U|$ is unchanged in (a) and receives very small effects in (b). These effects indicate that T_α is produced by the

D-state admixture with the linear dependence, $|S|$ is produced by the LS interaction, while T_β is produced by the second-order effect of the LS interaction. These results are consistent with those predicted in the previous section and provide justification for the considerations in Sec. 6.2.

Chapter 7

Models of ^{7}Li and Scattering
by Nuclei

Interactions of ^{7}Li with nuclei can be calculated by the folding model, for which we treat the ^{7}Li nucleus in two ways. One is the cluster model where the α-cluster (α) and the triton cluster (t) are bound to each other to form the ^{7}Li nucleus as shown in Fig. 7.1, while in the other model, called the continuum model, the ^{7}Li nucleus is considered as nuclear matter of a proper nucleon density. In the following, we will discuss the cluster model first and investigate the continuum model next. The scattering of ^{7}Li by nuclei is treated by the coupled-channel method, where the reaction channels for the excitations of ^{7}Li are taken into account.

7.1 General View of $\alpha + t$ Cluster Model

The ^{7}Li nucleus in the $\alpha + t$ cluster model resembles the deuteron, in the sense that both consist of two bodies, but ^{7}Li is very different from the deuteron in the interaction with the target nucleus in scattering. The deuteron has an almost spherical shape with small corrections due to the D-state admixture, which is characterized by a small electric quadrupole moment, $Q_e = 2.7$ mb. Thus, the T_R-type tensor interaction between the deuteron and the nucleus, which is derived from the D-state, is rather weak. On the other hand, ^{7}Li is greatly deformed, as is shown by the electric quadrupole moment, $Q_e = -36.5$ mb. The magnitude of Q_e is larger, when compared to that of the deuteron, by a factor of more than ten times. This large

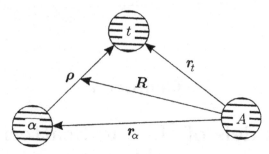

Fig. 7.1 $\alpha + t$ Cluster model in scattering of ^{7}Li. α, t and A stand for the α-cluster, the triton cluster and the target nucleus, respectively.

deformation of ^{7}Li yields a strong T_R-type tensor interaction. Further, the ^{7}Li-nucleus LS interaction is very weak, since the α-nucleus interaction has no LS part and the triton–nucleus LS interaction is weak when compared with that of the nucleon–nucleus LS one. In the deuteron case, the LS interaction produces the T_L-type tensor interaction, as discussed in the previous chapter. However, the weak LS interaction of ^{7}Li cannot produce such T_L-type tensor interactions.

Furthermore, in the deuteron, the excited states form a continuum spectrum, while in the ^{7}Li nucleus the lowest group of the excited states consist of three discrete levels, i.e. one bound state and two resonance states, as in Fig. 7.2. These states can be excited by the tensor interaction from the ground state, and in this sense, the ground state and these three states form a complete set in theoretical analyses. However, because of their discreteness, they produce an LS interaction of appreciable magnitude in the coupled-channel treatment. Such effects can be observed in the vector analyzing power in ^{7}Li-nucleus scattering.

7.2   Folding Interaction by the $\alpha + t$ Cluster Model

We will calculate the folding interaction between ^{7}Li and a structureless nucleus A by the $\alpha+t$ cluster model for ^{7}Li. We will consider two states for the $\alpha + t$ internal motion, the wave functions of which are described by $|\frac{1}{2}l; IM_I\rangle$ and $|\frac{1}{2}l'; I'M_I'\rangle$. They are the ground state or one of the excited states. Here, we denote the $\alpha+t$ relative coordinate and the associated orbital angular momentum by $\boldsymbol{\rho}$ and $\boldsymbol{l}$, and the

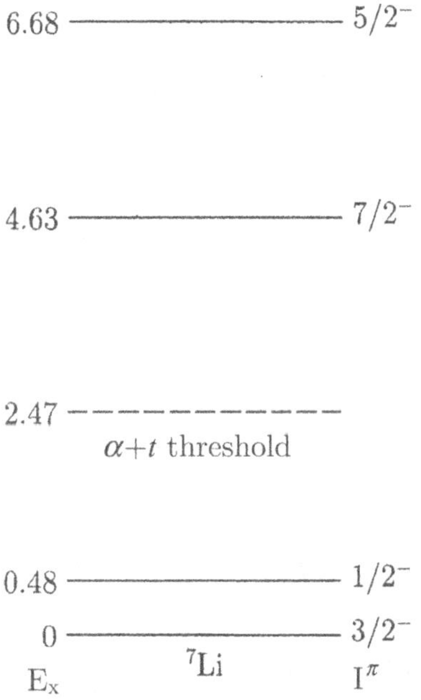

6.68 ———————————— 5/2⁻

4.63 ———————————— 7/2⁻

2.47 — — — — — — — — — —
       $\alpha+t$ threshold

0.48 ———————————— 1/2⁻
0 ———————————————— 3/2⁻
E_x ^7Li I^π

Fig. 7.2 Energy spectrum of ^7Li nucleus. $I^\pi = \frac{3}{2}^-$ and $\frac{1}{2}^-$ are the bound states and $I^\pi = \frac{7}{2}^-$ and $\frac{5}{2}^-$ are the resonance states. Excitation energy (E_x) is shown in MeV.

spin of ^7Li by I. Then $I = \frac{1}{2} + l$. Neglecting the triton-A LS interaction tentatively, the folding interaction is written as

$$\left\langle \frac{1}{2}l'; I'M_I' \middle| V \middle| \frac{1}{2}l; IM_I \right\rangle = \left\langle \frac{1}{2}l'; I'M_I' \middle| V_\alpha(r_\alpha) + V_t(r_t) \middle| \frac{1}{2}l; IM_I \right\rangle_\rho,$$

(7.1)

where V_α is the α-A interaction and V_t is the central part of the t-A interaction. The t-A and α-A relative coordinates, r_t and r_α (see Fig. 7.1), are transformed as

$$r_t = R + \frac{4}{7}\rho, \quad r_\alpha = R - \frac{3}{7}\rho,$$

(7.2)

where R is the coordinate of the center of mass of ^7Li relative to A. To perform the integration on ρ, we will expand $V_j(r_j)$ $(j = \alpha, t)$ in multipoles. Denoting $V_j(r_j)$ by $V_j(\boldsymbol{R}, \boldsymbol{\rho})$,

$$V_j(\boldsymbol{R}, \boldsymbol{\rho}) = \sum_\lambda V_j^{(\lambda)}(R, \rho)(C_\lambda(\hat{R}) \cdot C_\lambda(\hat{\rho})). \qquad (7.3)$$

The folding potential is given by the diagonal element of Eq. (7.1), where $|\frac{1}{2}l; IM_I\rangle$ is fixed to the ground state. After expanding the scalar product,

$$(C_\lambda(\hat{\rho}) \cdot C_\lambda(\hat{R})) = \sum_q (-)^q C_{\lambda-q}(\hat{\rho}) C_{\lambda q}(\hat{R}), \qquad (7.4)$$

we will apply the Wigner–Eckart theorem [3] to the matrix element of $C_{\lambda-q}(\hat{\rho})$

$$\left\langle \frac{1}{2}l; IM_I \left| C_{\lambda-q}(\hat{\rho}) \right| \frac{1}{2}l; IM_I \right\rangle$$

$$= \frac{1}{\sqrt{2I+1}}(I\lambda M - q \mid IM_I) \left(\frac{1}{2}l; I \left\| C_\lambda(\hat{\rho}) \right\| \frac{1}{2}l; I \right) \qquad (7.5)$$

and Ref. [3] gives

$$\left(\frac{1}{2}l; I \left\| C_\lambda(\hat{\rho}) \right\| \frac{1}{2}l; I \right)$$

$$= (-)^{\frac{1}{2}-l-\lambda-I}(2I+1)W(lIlI; \frac{1}{2}\lambda)(l\|C_\lambda(\hat{\rho})\|l), \qquad (7.6)$$

where $W(lIlI; \frac{1}{2}\lambda)$ is the Racah coefficient [3]. The Racah coefficient $W(j_1 j_2 J j_3; J_{12} J_{23})$ is defined as the transformation function from the representation of the addition of three vectors, $\boldsymbol{j}_1 + \boldsymbol{j}_2 = \boldsymbol{J}_{12}$, $\boldsymbol{J}_{12} + \boldsymbol{j}_3 = \boldsymbol{J}$ to that of another combination, $\boldsymbol{j}_2 + \boldsymbol{j}_3 = \boldsymbol{J}_{23}$, $\boldsymbol{j}_1 + \boldsymbol{J}_{23} = \boldsymbol{J}$. That is,

$$\langle j_1 j_2(J_{12})j_3; J \mid j_1, j_2 j_3(J_{23}); J \rangle$$

$$= \sqrt{(2J_{12}+1)(2J_{23}+1)}W(j_1 j_2 J j_3; J_{12} J_{23}). \qquad (7.7)$$

$(l\|C_\lambda\|l)$ in Eq. (7.6) is calculated as

$$(l\|C_\lambda\|l) = \sqrt{2l+1}(l\lambda 00 \mid l0), \qquad (7.8)$$

where $(l\lambda00 \mid l0)$ restricts λ to zero or even numbers. Using Eqs. (7.3)–(7.8), we obtain

$$\left\langle \frac{1}{2}l; IM_I \left| V_j(\boldsymbol{R}, \boldsymbol{\rho}) \right| \frac{1}{2}l; IM_I \right\rangle$$

$$= (-)^{l-I+\frac{1}{2}}\hat{l}\hat{I}\sum_\lambda (l\lambda00 \mid l0)W(lIlI; \frac{1}{2}\lambda)$$

$$\times \sum_q (-)^q(I\lambda M_I - q \mid IM_I)C_{\lambda q}(\hat{R})V_{lIlI,j}^{(\lambda)}(R) \qquad (7.9)$$

with

$$V_{lIlI,j}^{(\lambda)}(R) = \int u_{lI}^*(\rho)V_j^{(\lambda)}(R,\rho)u_{lI}(\rho)\rho^2 d\rho, \qquad (7.10)$$

where $u_{lI}(\rho)$ is the radial wave function of the $\alpha - t$ relative motion. The non-diagonal matrix element of Eq. (7.1) can be calculated similarly to the above derivation.

Next, we will show that the second line of Eq. (7.9) is equivalent to the matrix element of the T_R-type tensor operator of rank λ in the state $|IM_I\rangle$. When we extend the T_R tensor operator of rank 2 defined in Eq. (4.35) to rank λ,

$$T_R^{(\lambda)} = (S_\lambda(\boldsymbol{s}, \boldsymbol{s}) \cdot R_\lambda(\boldsymbol{R}, \boldsymbol{R})), \qquad (7.11)$$

where

$$S_{\lambda q}(\boldsymbol{s}, \boldsymbol{s}) = [\boldsymbol{s} \otimes \boldsymbol{s}]_q^\lambda, \quad R_{\lambda q}(\boldsymbol{R}, \boldsymbol{R}) = [\boldsymbol{R} \otimes \boldsymbol{R}]_q^\lambda. \qquad (7.12)$$

The matrix element in the state $|IM_I\rangle$ is given as

$$\langle IM_I| \sum_q (-)^q S_{\lambda-q}(\boldsymbol{s}, \boldsymbol{s}) \cdot R_{\lambda q}(\boldsymbol{R}, \boldsymbol{R})|IM_I\rangle$$

$$= \sum_q (-)^q \langle IM_I|S_{\lambda-q}(\boldsymbol{s}, \boldsymbol{s})|IM_I\rangle R_{\lambda q}(\boldsymbol{R}, \boldsymbol{R})$$

$$= \sum_q (-)^q(I\lambda M_I - q \mid IM_I)\frac{1}{\hat{I}}(I\|S^\lambda\|I)R_{\lambda q}(\boldsymbol{R}, \boldsymbol{R}).$$

$$(7.13)$$

The last line is obtained by applying the Wigner–Eckart theorem to the matrix element of $S_{\lambda-q}$. One can see that the second line of Eq. (7.9) is equivalent to Eq. (7.13), when the strength of $T_R^{(\lambda)}$ is properly adjusted, since $R_{\lambda q}(\boldsymbol{R}, \boldsymbol{R}) \propto C_{\lambda q}(\hat{R})$ as in the case of $\lambda = 2$. The above development means that the folding potential equation (7.1) can be expanded into T_R tensor terms of the rank λ. However, the value of λ is restricted by $0 \leq \lambda \leq 2I$ due to the CG factor $(I\lambda M_I - q \mid IM_I)$. The ^{7}Li-$A$ interaction obtained by folding the α-A and t-A central potentials consists of the central interaction ($\lambda = 0$) and the second-rank tensor one ($\lambda = 2$), since $\lambda \leq 3$ due to $I = \frac{3}{2}$ and λ must either be zero or an even number. The magnitudes of these interactions depend on the $\alpha - t$ wave function $u_{lI}(\rho)$ through $V_{lIlI,j}^{(\lambda)}(R)$.

Finally, we will examine the contribution of the triton spin-orbit interaction to the ^{7}Li-nucleus spin-orbit one in the folding model. The change of variables from $(\boldsymbol{r}_t, \boldsymbol{r}_d)$ to $(\boldsymbol{\rho}, \boldsymbol{R})$, Eq. (7.2), gives $\boldsymbol{\nabla}_t = \frac{3}{7}\boldsymbol{\nabla}_R + \boldsymbol{\nabla}_\rho$. Then we get

$$
l_t = \frac{\hbar}{i}\boldsymbol{r}_t \times \boldsymbol{\nabla}_t = \frac{\hbar}{i}\left(\boldsymbol{R} + \frac{4}{7}\boldsymbol{\rho}\right) \times \left(\frac{3}{7}\boldsymbol{\nabla}_R + \boldsymbol{\nabla}_p\right)
$$

$$
= \frac{\hbar}{i}\left\{\frac{3}{7}\boldsymbol{R} \times \boldsymbol{\nabla}_R + \frac{12}{49}\boldsymbol{\rho} \times \boldsymbol{\nabla}_R + \boldsymbol{R} \times \boldsymbol{\nabla}_\rho + \frac{4}{7}\boldsymbol{\rho} \times \boldsymbol{\nabla}_\rho\right\}. \quad (7.14)
$$

The contribution to the ^{7}Li-nucleus orbital angular momentum, $\boldsymbol{L} = \frac{\hbar}{i}\boldsymbol{R} \times \boldsymbol{\nabla}_R$, arises from the first term of the right-hand side of the above equation. Using

$$
s_t = \boldsymbol{I} - \boldsymbol{l} \quad (7.15)
$$

and the multipole expansion of $V_{so}(r_t)$

$$
V_{so}(r_t) = \sum_\lambda V_{so}^{(\lambda)}(R, \rho)(C_\lambda(\hat{\rho}) \cdot C_\lambda(\hat{R})), \quad (7.16)
$$

we obtain the ^{7}Li-nucleus spin-orbit interaction as

$\boldsymbol{I} \cdot \boldsymbol{L}$ component of $\langle s_t \cdot l_t V_{so}(r_t)\rangle_{\text{folding}}$

$$
= \frac{3}{7}\boldsymbol{I} \cdot \boldsymbol{L}\int V_{so}^{(0)}(R, \rho)|u_{l=1\ I=\frac{3}{2}}(\rho)|^2 \rho^2 d\rho, \quad (7.17)
$$

where $u_{l=1\,I=\frac{3}{2}}(\rho)$ is the radial part of the ^{7}Li internal wave function in the ground state.

7.3 Numerical Calculation of Scattering by Cluster Model and Comparison with Experimental Data at $E_{lab} = 20.3$ MeV

Numerical calculations are performed for the ^{7}Li-^{58}Ni scattering at $E_{lab} = 20.3$ MeV, by the use of the folding interaction. To obtain the folding interaction we need two kinds of interactions, i.e. the one for the α-t relative motion in ^{7}Li and the other for the t-^{58}Ni and α-^{58}Ni interactions in scattering states. For the latter, we assume optical potentials, which include imaginary parts. For these interactions, we will adopt the Woods–Saxon type form factors.

For the α-t interaction, the potential parameters are fixed for each partial wave so as to reproduce the scattering phase shifts at low energies. The parameters obtained are given in Table 7.1 and the calculated phase shifts are compared with the empirical values in Fig. 7.3, showing reasonable agreement. The calculation also reproduces the data of the binding energy and the quadrupole moment for the ground state, and those of the binding energy and the $B(E2)$

Table 7.1 α-t potential parameters where the unit of length (energy) is fm (MeV).

l	J^π	V_O	V_{SO}
0	$\frac{1}{2}^+$	-76.0	
1	$\frac{1}{2}^-, \frac{3}{2}^-$	-93.0	-2.0
3	$\frac{5}{2}^-, \frac{7}{2}^-$	-90.5	-3.5

$W = W_D = 0,$

$r_o = r_c = 2.05, \ a_0 = 0.70.$

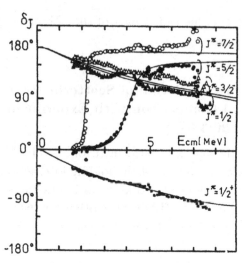

Fig. 7.3 Comparison of calculated phase shifts with the empirical values for $J^\pi = \frac{1}{2}^+, \frac{1}{2}^-, \frac{3}{2}^-, \frac{5}{2}^-$ and $\frac{7}{2}^-$ partial waves at E_{cm} up to 7 MeV.

Table 7.2 Parameters for t-^{58}Ni and α-^{58}Ni optical potentials. The unit of length (energy) is fm (MeV).

	V	W_V	V_{so}	r_V	r_W	r_{so}	r_c	a_V	a_W	a_{so}
t-^{58}Ni	-151.0	-18.2	-4.0	1.20	1.60	1.10	1.30	0.66	0.83	0.83
α-^{58}Ni	-169.5	-18.9		1.43	1.43		1.40	0.50	0.50	

strength for the $\frac{1}{2}^-$ excited state [26]. These potentials supply the α-t relative wave functions $u_{lI}(\rho)$.

As discussed in Chapter 5, the optical potential is energy-dependent. At present, the energies of the α-particle and the triton are determined by sharing the ^{7}Li energy among the α-particle and the triton proportionally to their masses, so that they have the same velocity. Taking this condition into account, the potential parameters are determined by extrapolations from those at higher energies. The parameters adopted are listed in Table 7.2. The obtained ^{7}Li-^{58}Ni interactions are displayed numerically in Fig. 7.4 [26]. In the

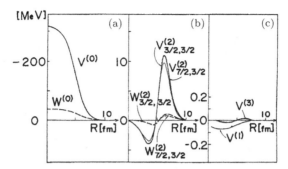

Fig. 7.4 Folding interactions between ^{7}Li and ^{58}Ni. $V^{(0)}$ and $W^{(0)}$ are the real and imaginary parts for the central interaction and $V^{(2)}$ and $W^{(2)}$ are for those of the second-rank tensor ones. The suffixes I and I' indicate the initial and final states connected by the interaction. $V^{(1)}$ and $V^{(3)}$ are the LS and third-rank tensor interactions, respectively.

figure, $V^{(0)}$ and $W^{(0)}$ are the form factors of the real and imaginary parts of the central interaction, and $V^{(2)}_{I'I}$ and $W^{(2)}_{I'I}$ are those of the second-rank tensor interactions. In this latter case, the suffixes I and I' indicate the initial and final states in the excitation produced by the interaction, e.g. $V^{(2)}_{7/2,3/2}$ stands for the tensor interaction for the excitation from the ground state of $I = \frac{3}{2}$ to the $I = \frac{7}{2}$ excited state. $V^{(2)}_{1/2,3/2}$ is very similar to $V^{(2)}_{3/2,3/2}$ and they can hardly be distinguished from each other. A similar situation is seen in $V^{(2)}_{5/2,3/2}$ and $V^{(2)}_{7/2,3/2}$. Finally, it should be emphasized that the calculated $V^{(1)}$, which is the form factor of the LS interaction, has too small a magnitude to explain the observed vector analyzing power iT_{11}. Also, the form factor of the third-rank tensor interaction $V^{(3)}$, which is constructed by the LS interaction and the second-rank tensor one, is very weak.

The comparison of the calculated observables [26] with the measured ones [27] is shown in Fig. 7.5, where the Rutherford ratio of the cross section σ/σ_R, the vector analyzing power iT_{11} and the tensor analyzing powers T_{20}, T_{21} and T_{22} in the ^{7}Li $+$ ^{58}Ni scattering at $E_{\text{lab}} = 20.3$ MeV are investigated. In the figure, the dotted lines describe the contributions of the ^{7}Li ground state, which are calculated through the use of the folding interactions with their form

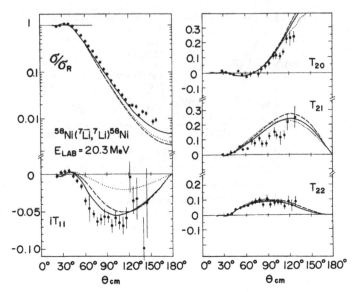

Fig. 7.5 Comparison between calculation and experiment in the Rutherford ratio of the cross section $\frac{\sigma}{\sigma_R}$, the vector analyzing power iT_{11} and the tensor analyzing powers T_{20}, T_{21} and T_{22} for the ^{7}Li-^{58}Ni elastic scattering at $E_{\text{lab}} =$ 20.3 MeV. The dotted lines are for the ground-state contribution. The dashed and solid lines include the virtual excitations (see text).

factors in Fig. 7.4, but the couplings to the ^{7}Li excited states are ignored. The dashed and solid lines take into account the contributions of the virtual excitations of the excited states, in addition to those of the ground state, by the coupled channel method with the coupling form factors in Fig. 7.4. In detail, as the excited state, the dashed lines include the $\frac{1}{2}^-$ bound state and the solid lines include the $\frac{1}{2}^-$ state and the $\frac{5}{2}^-$ and $\frac{7}{2}^-$ resonance states. The contribution of the $\frac{5}{2}^-$ state is very small at the present energy.

In the calculation of the vector analyzing power, the contribution of the vector interaction $V^{(1)}$ is negligibly small, but the higher order of the tensor interaction produces the vector amplitude in the ground state. Further, in the coupled-channel calculation, the tensor interaction produces additional vector amplitudes through the virtual excitations of the excited states. These effects of the tensor interaction explain the experimental data of the vector anlyzing power. At the present energy, the contributions of the $\frac{1}{2}^-$ and $\frac{7}{2}^-$ excited states are remarkable.

On the cross section, the ground-state contribution reproduces the experimental data at small angles but underestimates the magnitude at large angles. However, the virtual excitation of the $\frac{7}{2}^-$ state increases the large-angle cross section and gives satisfactory agreement with the data. The measured tensor analyzing powers are well reproduced by the calculation, where the ground-state contribution is dominant and the virtual excitations to the ^7Li excited states provide relatively small corrections.

Finally, it is noted that, in the present case, the adiabatic approximation on the Green's function in the two-step approach discussed in the previous chapter will lead to serious errors, because the states included in the channel coupling calculation have their own value of the energy and cannot be replaced by the average value. To emphasize the situation, a coupled-channel calculation of iT_{11} is displayed in Fig. 7.6, where all excited states are lowered hypothetically to the ground state and the channel coupling is considered only between each excited state and the ground state. The contribution of the $\frac{7}{2}^-$ state cancels those of the ground state and the $\frac{1}{2}^-$ excited

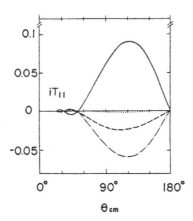

Fig. 7.6 iT_{11} by the calculation of the adiabatic approximation. The dash-dotted, dotted and solid lines are for the $\frac{1}{2}^-$, $\frac{5}{2}^-$ and $\frac{7}{2}^-$ excited states. The dashed line is for the contribution of the ground state.

state and the resultant iT_{11} vanishes. This is completely different from the realistic calculation in Fig. 7.5. These characteristics of the virtual-excitation effects are clarified analytically for the general case, though manipulations are discarded at present because of certain complications [26].

7.4 Folding Interaction by Continuum Model and Comparison with Experimental Data in Scattering by ^{120}Sn at $E_{\text{lab}} = 44$ MeV

In this section, we will derive the ^{7}Li-target nucleus folding interaction with the continuum model for ^{7}Li [28]. The interactions obtained are examined by comparing the calculated observables with the measured ones. The interaction is classified in the following two ways according to the nuclear model employed for the target nucleus. These are the single folding and double folding interactions, as illustrated in Fig. 7.7. In the single folding model, the target nucleus is assumed to be a structure-less single body, and the ^{7}Li-nucleus

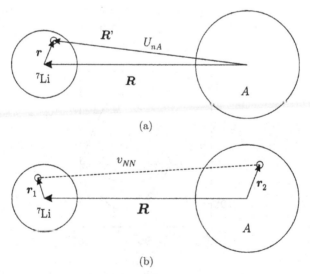

(a)

(b)

Fig. 7.7 Folding interaction for the ^{7}Li continuum model. (a) Single folding, (b) double folding.

interaction is obtained by folding the optical potential of the nucleon of ^{7}Li from the target nucleus, by the use of the nucleon density of ^{7}Li. In the double folding model, the target nucleus is assumed to have a continuum structure, and the ^{7}Li-nucleus interaction is obtained by folding an effective nucleon–nucleon interaction between the nucleons of ^{7}Li and of the target, by the use of the nucleon density of ^{7}Li and that of the target nucleus. In both cases of folding, the formulae of the folding interaction can also be applied to the inelastic scattering which excites ^{7}Li, for which the transition density from the ground state to the excited state is applied for ^{7}Li.

The interaction is denoted by V_{ij}, where i stands for the ^{7}Li ground state and j for the ground state, or one of the excited states. The single folding (hereafter abbreviated by SF) interaction $V_{ij}^{SF}(R)$ is given by

$$V_{ij}^{SF}(\boldsymbol{R}) = \int U_{nA}(\boldsymbol{R}')\rho_{ij}^{(P)}(r)d\boldsymbol{r} \qquad (7.18)$$

with

$$\boldsymbol{R}' = \boldsymbol{R} + \boldsymbol{r}, \qquad (7.19)$$

where U_{nA} is the optical potential and $\rho_{ij}^{(P)}$ is the nucleon density for $j = 1$ and the transition density for $j \neq i$ in ^{7}Li. In practice, $V_{ij}^{SF}(\boldsymbol{R})$ consists of a central interaction and a tensor interaction. The tensor part is produced by the deformation of $\rho_{ij}^{(P)}(\boldsymbol{r})$. The spin-orbit part of V_{ij}^{SF} is ignored because most of the spin-orbit part of U_{nA} is considered to vanish in the folding procedure.

The double folding (hereafter abbreviated by DF) interaction $V_{ij}^{DF}(\boldsymbol{R})$ is given by

$$V_{ij}^{DF}(\boldsymbol{R}) = (1 + iN_I) \int \rho_{ij}^{(P)}(\boldsymbol{r}_1)\rho_{00}^{(T)}(\boldsymbol{r}_2)v_{NN}(\boldsymbol{r}_1 + \boldsymbol{R} - \boldsymbol{r}_2)d\boldsymbol{r}_1 d\boldsymbol{r}_2, \qquad (7.20)$$

where $\rho_{00}^{(T)}$ is the nucleon density of the target nucleus and v_{NN} is the effective nucleon–nucleon interaction, for which we assume the so-called $M3Y$ interaction [29] with $r = r_1 + R - r_2$

$$v_{NN} = \left[7999\frac{e^{-4r}}{4r} - 2134\frac{e^{-2.5r}}{2.5r} + \hat{J}_{00}\delta(r) \right] \text{ MeV}, \qquad (7.21)$$

where $\hat{J}_{00} = -262\,\text{fm}^3$ for $E = 10$ MeV. Since v_{NN} has no imaginary part, to describe the absorption of the beam, we have included an imaginary part N_I in V_{ij}^{DF}, which will be treated as a flexible parameter in numerical analyses.

The ^{7}Li densities $\rho_{ij}^{(P)}$ are calculated by the microscopic wave functions of ^{7}Li, which are obtained by introducing the internal wave functions of the α-particle and the triton for the $\alpha + t$ cluster model of ^{7}Li, where the antisymmetrization among the seven nucleons is taken into account. The measured binding energy and quadrupole moment for the ground state, the excitation energy and $B(E2)$ for the $\frac{1}{2}^-$ excited state, and the phase shifts for the α-t scattering, are reproduced as in the preceding section [32].

Now we will examine the SF and DF interactions by comparing the calculated observables with the experimental data, together with the values calculated from the cluster folding (abbreviated as CF) interaction in the previous sections. The comparison will be made on the scattering by the ^{120}Sn target at $E_{\text{lab}} = 44$ MeV [30] for both elastic and inelastic scattering to the $\frac{1}{2}^-$ state of ^{7}Li [31]. As the observables, we treat the Rutherford ratio of the cross section σ/σ_R for the elastic scattering and the cross section σ for the inelastic scattering, the vector analyzing power iT_{11}, the tensor analyzing powers, $^TT_{20}$, T_{20} and T_{21}, and the third-rank tensor analyzing power $^TT_{30}$, where $^TT_{20}$ and $^TT_{30}$ are measured as analyzing powers for the aligned beam discussed in Sec. 3.4. They are connected to the analyzing powers of the polarized beams as

$$^TT_{20} = -\frac{1}{2}(T_{20} + \sqrt{6}T_{22}) \qquad (7.22)$$

and

$$^TT_{30} = -\frac{1}{2}(\sqrt{3}iT_{30} + \sqrt{5}iT_{33}). \qquad (7.23)$$

As will be seen later, the calculation closely reproduces the data of the analyzing powers, but not those of the elastic cross section at large angles. To improve the agreement in the cross section, we will introduce the normalization factors N_R and N_I for the real and imaginary parts of each interaction. For the CF interaction, $(N_R, N_I) = (0.62, 0.61)$. In the case of the SF interaction, we will examine two sets of nucleon optical potentials as the input: one by Wilmore and Hodgson (abbreviated as WH) and the other by Becchetti and Greenlees (abbreviated as BG). For the WH case, the following values are used, $(N_R, N_I) = (0.29, 0.675)$, and for the BG case, $(N_R, N_I) = (0.15, 0.185)$ are used. For the DF potential, no normalization is applied for the real part, then $(N_R, N_I) = (1.0, 0.45)$ is employed. The comparison of the calculated observables with the measured ones are shown in Fig. 7.8 for the CF and DF models, and in Fig. 7.9 for the SF model with the BG potential and with the WH potential. The calculations are performed in the coupled-channel framework, which includes, for ^7Li, the $\frac{1}{2}^-$, $\frac{7}{2}^-$ and $\frac{5}{2}^-$ excited states in addition to the $\frac{3}{2}^-$ ground state. Thus, the effects of the virtual excitation to these excited states are taken into account.

The calculations by the CF, DF and SF models provide similar magnitudes and angular dependences for each observable, in spite of the many kinds of observables treated. Also, the calculated quantities reproduce the global feature of the measured quantities, except for $^T T_{20}$ and T_{21} by the DF model for the elastic scattering at large angles. Thus, agreement with the data is almost independent of the models used in both the elastic and the inelastic scattering.

To investigate the origin of such specific features of the calculation, we will display, in Fig. 7.10, the real and imaginary parts of the ground-state potential V_{00} as functions of R. There, the real and imaginary parts of V_{00} used in the various models are concentrated at a particular R. $R_r = 11.2$ fm for $\mathrm{Re}(V_{00})$ and $R_i = 10.0$ fm for $\mathrm{Im}(V_{00})$. This suggests the interaction in a limited region around R_r and R_i to be effective for the scattering. The large magnitudes of R_r and R_i can be interpreted as the result of the scattering, which happens far from the nuclear surface due to the large size of ^7Li.

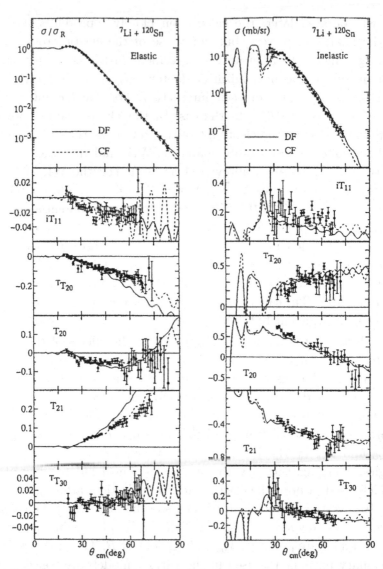

Fig. 7.8 Comparison of the calculated observables using the CF and DF models with the experimental data. The normalization factors of the interaction are $(N_R, N_I) = (0.62, 0.61)$ for the CF model and $(1.0, 0.45)$ for the DF model. The observables measured are the cross section σ (Rutherford ratio σ/σ_R for the elastic scattering), the vector analyzing power iT_{11}, the tensor analyzing powers, $^T T_{20}$, T_{20}, T_{21} and the third-rank tensor analyzing power $^T T_{30}$ for the elastic scattering of ^{7}Li and the inelastic scattering to the $\frac{1}{2}^-$ excited state of ^{7}Li by the ^{120}Sn target at $E_{\text{lab}} = 44$ MeV.

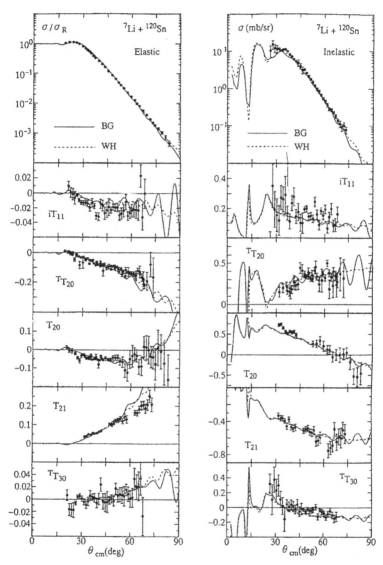

Fig. 7.9   Comparison of the calculated observables using the SF model with the experimental data. The BG and WH optical potentials are used as the input. The normalization factors of the interaction are $(N_R, N_I) = (0.25, 0.185)$ for the BG potential, and $= (0.29, 0.675)$ for the WH potential. For further information, see the caption of Fig. 7.8.

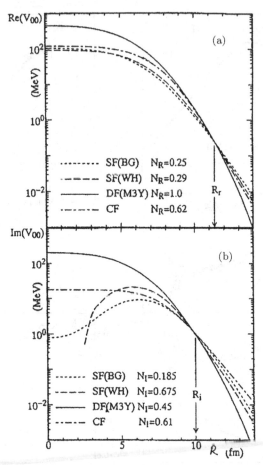

Fig. 7.10   Form factors of the central interaction for the SF, DF and CF models. (a) Re and (b) Im refer to the real and imaginary parts of the interaction respectively.

7.5   Characteristics of the Contribution of the Tensor Interaction

In the preceding sections, we found that the ^{7}Li-nucleus spin-dependent interaction is characterized by a strong tensor interaction, which is induced by a large deformation of ^{7}Li. In that case, polarization observables of ^{7}Li-nucleus scattering will be sensitive to variations in this interaction. To examine such features of the tensor interaction quantitatively, we will now compare the observables obtained by the standard calculation, where the strength of the tensor interaction is realistic, with those obtained by a hypothetical calculation, where the strength of the tensor interaction is artificially reduced by the factor $\frac{1}{2}$. The comparison is based on the coupled-channel calculation for the ^{120}Sn target at $E_{\text{lab}} = 44$ MeV. For that calculation, we choose the CF model for the ^{7}Li-nucleus interaction, but without the normalization of the interaction strength. Although the comparison with the experimental data is not shown, the standard calculation accurately reproduces the data, except for the elastic cross section at large angles.

A comparison of the two kinds of calculations described above is shown in Fig. 7.11, where, in the elastic scattering, the tensor analyzing powers T_{20}, T_{21} and $^T T_{20}$ are reduced by a factor of about $\frac{1}{2}$ by the reduction of the tensor interaction strength without changes in their angular distributions. At the same time, the cross section is almost unchanged except at large angles. On the other hand, in the inelastic scattering, the tensor analyzing powers are almost unchanged, while the cross section is reduced by a factor of about $\frac{1}{4}$. These characteristics of the calculation can be understood from the constitution of the invariant amplitudes of the observables.

The observables by the invariant amplitudes are given for the elastic scattering in Sec. 4.4. They are described by the amplitudes of the scalar U, the vector S, the tensors $T_{2\alpha}$ and $T_{2\beta}$, and the third-rank tensors $T_{3\alpha}$ and $T_{3\beta}$. For the inelastic scattering, the observables are given in terms of the amplitudes $A \sim D$, which can be transformed to the invariant amplitudes of the vector S and the tensors $T_{2\alpha}$, $T_{2\beta}$ and $T_{2\gamma}$. To simplify the consideration, here we will introduce the following approximation. Elastic scattering is strongly induced by the central interaction, the amplitude of which is given by U. We will, therefore, keep in the observables only the terms which include U. Thus, we get for the elastic scattering

$$\sigma = \left(\frac{\mu_{aA}}{2\pi}\right)^2 \frac{1}{A} N_R \quad \text{with} \quad N_R = |U|^2, \tag{7.24}$$

$$iT_{11} = \frac{2\sqrt{2}}{\sqrt{5}\, N_R} \text{Im}\{US^*\}, \tag{7.25}$$

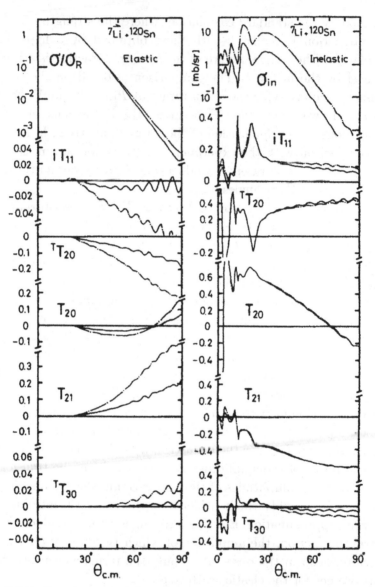

Fig. 7.11 Contributions of the tensor interaction in the elastic scattering and the inelastic scattering to the excited state of ^{7}Li ($\frac{1}{2}^-$) by the ^{120}Sn target at $E_{\text{lab}} = 44$ MeV. The dash-dotted lines show the calculation by the CF model and the solid lines show the hypothetical calculation where the strength of the tensor interaction is reduced by the factor $\frac{1}{2}$. For further information, see the caption of Fig. 7.8.

$$T_{20} = \frac{2}{N_R} \mathrm{Re}\{UT_{2\alpha}^*\}, \tag{7.26}$$

$$T_{21} = \frac{1}{\sqrt{2}\,N_R} \mathrm{Re}\{U(-\sqrt{3}\,T_{2\alpha} + T_{2\beta})^*\}\tan\theta, \tag{7.27}$$

$$^TT_{20} = -\frac{1}{N_R}\mathrm{Re}\{U(T_{2\alpha} + \sqrt{3}\,T_{2\beta})^*\} \tag{7.28}$$

and

$$^TT_{30} = \frac{1}{\sqrt{5}\,N_R}\mathrm{Im}\{U(-2\sqrt{3}\,T_{3\alpha}\cot\theta - 8T_{3\beta})^*\}. \tag{7.29}$$

These equations indicate that in the elastic scattering the cross section is not affected by the reduction of the tensor interaction strength, while the tensor analyzing powers are reduced in proportion to the strength of the tensor interaction. This prediction is justified by the numerical calculation shown in Fig. 7.11, which compares the realistic calculation with the hypothetical calculation. In a detailed examination of the figure, the cross section at large angles is affected by the reduction of the tensor interaction. This will be the effects of tensor terms in the cross section which were neglected in Eq. (7.24). As discussed in Sec. 7.2, the second-order of the tensor interaction produces the vector amplitudes S. Then, Eq. (7.25) suggests the calculated iT_{11} is decreased by the factor $\frac{1}{4}$ if the tensor interaction strength is reduced by $\frac{1}{2}$. The large reduction of $^TT_{30}$ indicates that the third-rank tensor amplitudes, $T_{3\alpha}$ and $T_{3\beta}$, are produced by higher orders of the second-rank tensor interaction.

In the inelastic scattering to the $\frac{1}{2}^-$ state of ^7Li, the excitation of ^7Li is induced by the second-rank tensor interaction. Thus, it would be a reasonable approximation to keep only the terms of $T_{2\alpha}$, $T_{2\beta}$ and $T_{2\gamma}$. In this approximation, for the inelastic scattering we get

$$\sigma = \left(\frac{\mu_{aA}}{2\pi}\right)^2 \frac{k_f}{k_i}\frac{N_R}{4} \quad \text{with} \quad N_R = \frac{1}{2}|T_{2\alpha}|^2 + 2|T_{2\beta}|^2 + 2|T_{2\gamma}|^2, \tag{7.30}$$

$$iT_{11} = \frac{1}{\sqrt{10}\,N_R}\mathrm{Im}\{-3\sqrt{3}\,T_{2\alpha}T_{2\beta}^* + 3T_{2\alpha}T_{2\gamma}^*\}, \tag{7.31}$$

$$T_{20} = \frac{2}{N_R}\left(-\frac{1}{8}|T_{2\alpha}|^2 - |T_{2\beta}|^2 + |T_{2\gamma}|^2\right), \qquad (7.32)$$

$$T_{21} = \frac{1}{\sqrt{2}\,N_R}\mathrm{Re}\{T_{2\alpha}T_{2\beta}^* + \sqrt{3}T_{2\alpha}T_{2\gamma}^*\}, \qquad (7.33)$$

$$^T T_{20} = -\frac{1}{2N_R}\{-|T_{2\alpha}|^2 - 2|T_{2\beta}|^2 + 2|T_{2\gamma}|^2 + 4\sqrt{3}\,\mathrm{Re}(T_{2\beta}T_{2\gamma}^*)\},$$

$$(7.34)$$

$$^T T_{30} = -\frac{1}{\sqrt{5}\,N_R}\mathrm{Im}\{\sqrt{3}T_{2\alpha}T_{2\beta}^* - T_{2\alpha}T_{2\gamma}^*\}. \qquad (7.35)$$

In Eq. (7.30), one can see that the cross section consists of the terms of absolute square of the tensor amplitude. This explains why the reduction of the cross section happens by the factor $\frac{1}{4}$ when the reduction of the tensor interaction strength occurs by the factor $\frac{1}{2}$. Contrary to this, in the analyzing powers, Eqs. (7.31)–(7.35), the effects of the reduction of the tensor interaction strength in the numerator will be cancelled by those in the denominator. Then the analyzing powers will not be affected by the reduction of the tensor interaction strength. In Fig. (7.11), some reduction effects are seen in iT_{11} and $^T T_{30}$ at $\theta \gtrsim 40°$, which will be due to the contribution of the vector amplitudes. This is ignored in Eqs. (7.31) and (7.35).

Chapter 8

Polarization in Resonance Reaction

Theoretically, polarization in nuclear reactions is induced by the modification of the transition amplitude by spin-dependent interactions. Accordingly, the study of polarization plays an important role in the examination of the spin-dependent interaction. Resonance and polarization will be related to each other in a similar manner to the above relation between a spin-dependent interaction and polarization.

In resonance reactions, the transition amplitude is restricted by the spin of the resonance state concerned. For example, in pure resonance reactions, only the amplitude which has the same total angular momentum as the resonance spin is effective. Such requirements on the transition amplitude will induce polarization of the related particles. Conversely, studies of polarization offer a decisive means for the determination of the resonance spin.

8.1 How to Derive Spin Parity of Resonance from Analyzing Power Data

In this chapter, we will consider the deuteron tensor analyzing powers T_{2q} ($q = 0, 1, 2$) of $^6\mathrm{Li}(d, p)\,^7\mathrm{Li}(\frac{3}{2}^-\ g.s.)$ reactions at $E_d = 90\,\mathrm{keV}$. The $d + {}^6\mathrm{Li}$ system at this incident energy corresponds to the excited state of $^8\mathrm{Be}$ at $E_{\mathrm{ex}} = 22.2\,\mathrm{MeV}$, around which a broad resonance has been expected from the phase-shift analysis of the $\alpha + \alpha$ scattering. The resonance is characterized by $E_{\mathrm{res}} = 22.09\,\mathrm{MeV}$ and $\Gamma = 580\,\mathrm{keV}$. We will try to determine the spin parity of this resonance by investigating experimental data of the deuteron analyzing powers T_{2q}.

The analysis will be performed in two ways. First, we will analyze the reaction in a model-independent way by using the invariant amplitudes. The amplitudes are chosen consistently with a given spin parity, from which the analyzing powers are calculated. The comparison of the calculated analyzing powers with the data determines the propriety of the spin parity assumption. Next, we adopt the DWBA reaction model. The zero-range assumption is employed for the effective interaction. In this treatment, the calculation is performed explicitly under the assumption that the reaction takes place through a resonance of the given spin parity. The comparison of the calculated analyzing power with the measured one determines the spin parity of the resonance. Finally, the results of the two different approaches are found to be consistent with each other.

8.2 Tensor Analyzing Powers by the Invariant Amplitude Method

The invariant amplitude method developed in Chap. 4 presents the general formula of the transition amplitude for nuclear reactions. For a $A(d,p)B$ reaction, the transition amplitude is described as

$$\langle \nu_p \nu_B; \boldsymbol{k}_f | M | \nu_d \nu_A; \boldsymbol{k}_i \rangle = \sum_K \sum_{s_i s_f} (s_d s_A \nu_d \nu_A \mid s_i \nu_i)(s_p s_B \nu_p \nu_B \mid s_f \nu_f)$$

$$\times \langle s_f \nu_f; \boldsymbol{k}_f | M^{(K)} | s_i \nu_i; \boldsymbol{k}_i \rangle, \qquad (8.1)$$

where M is the T-matrix of the reaction, $s(\nu)$ stands for spins (z components) and $\boldsymbol{k}$'s for relative momenta. The partial amplitude $\langle s_f \nu_f; \boldsymbol{k}_f | M^{(K)} | s_i \nu_i; \boldsymbol{k}_i \rangle$ describes the transition induced by the interaction of the tensor rank K and is given by

$$\langle s_f \nu_f; \boldsymbol{k}_f | M^{(K)} | s_i \nu_i; \boldsymbol{k}_i \rangle$$

$$= (-)^{s_f - \nu_f} (s_i s_f \nu_i - \nu_f \mid K\kappa)$$

$$\times \sum_{l_i = \bar{K} - K}^{K} [C_{l_i}(\hat{k}_i) \otimes C_{l_f = \bar{K} - l_i}(\hat{k}_f)]_\kappa^K F(s_i s_f K l_i). \qquad (8.2)$$

Here $F(s_i s_f K l_i)$ is the invariant amplitude, where the dependencies on E and $\cos\theta$ are discarded for simplicity. C_{lm} is given by

$$C_{lm} = \sqrt{\frac{4\pi}{2l+1}} Y_{lm}. \tag{8.3}$$

Because of the very low incident energy, the S-wave will be sufficient to describe the incident wave. Contributions of higher partial waves will be discussed later. Then we will set $l_i = 0$ and treat the invariant amplitudes $F(s_i s_f K l_i)$ as being θ-independent. From Eq. (8.2), $l_i + l_f = K$, which gives $K = l_f$. Due to the parity conservation, l_f has odd parity since the neutron is transferred from the S-state of the deuteron to the P-state of the residual nucleus B. In Eq. (8.1), $s_d + s_A = s_i$ and $s_d = s_A = 1$ give $s_i = 0, 1, 2$. Similarly $s_p + s_B = s_f$ and $s_p = \frac{1}{2}$, $s_B = \frac{3}{2}$ give $s_f = 1, 2$. Then from Eq. (8.2), $K = 0, 1, 2, 3, 4$, among which $K = 1$ and 3 are allowed because K is odd. Further, choosing the Madison convention for the coordinate system, $z \parallel k_i$ and $y \parallel k_i \times k_f$, and we get

$$\sum_{l_i} [C_{l_i}(\hat{k}_i) \otimes C_{l_f}(\hat{k}_f)]^K_\kappa = \sqrt{\frac{4\pi}{2K+1}} Y_{K\kappa}(\hat{k}_f). \tag{8.4}$$

The cross section $\frac{d\sigma}{d\Omega}$ and the tensor analyzing powers T_{2q} are given in Chap. 3,

$$\frac{d\sigma}{d\Omega} = \frac{1}{(2s_d+1)(2s_A+1)} N_R \quad \text{with} \quad N_R = Tr(MM^\dagger) \tag{8.5}$$

and

$$T_{2q} = \frac{1}{N_R} Tr(M\tau_q^2 M^\dagger), \tag{8.6}$$

where the matrix element of τ_q^2 is given by Eqs. (2.6) and (2.7)

$$\langle s_d \nu_d | \tau_q^2 | s_d \nu_d' \rangle = (s_d 2 \nu_d' q \mid s_d \nu_d) \sqrt{5}. \tag{8.7}$$

Using the above, we can write the contribution of a particular K term to the analyzing power as

$$
\begin{aligned}
N_R T_{2q}(K) = \sum_{\nu_p \nu_B} \sum_{\nu_d \nu_A} \sum_{s_i s_f} & (s_d s_A \nu_d \nu_A \mid s_i \nu_i)(s_p s_B \nu_p \nu_B \mid s_f \nu_f) \\
& \times \langle s_f \nu_f; \boldsymbol{k}_f | M^{(K)} | s_i \nu_i; \boldsymbol{k}_i \rangle (s_d 2 \nu'_d q \mid s_d \nu_d) \\
& \times \sum_{s'_i s'_f} (s_d s_A \nu'_d \nu_A \mid s'_i \nu'_i)(s_p s_B \nu'_p \nu_B \mid s'_f \nu'_f) \\
& \times (s'_f \nu'_f; \boldsymbol{k}_f | M^{(K)} | s'_i \nu'_i; \boldsymbol{k}_i)^* \sqrt{5}.
\end{aligned} \tag{8.8}
$$

The calculation is carried out by inserting Eqs. (8.2) with (8.4) into (8.8). To perform the operations $\sum_{\nu_p \nu_B} \sum_{\nu_d \nu_A}$, we will transform the CG coefficients. For example, the following one is employed for $\sum_{\nu_A}$,

$$
\begin{aligned}
(s_d s_A \nu'_d \nu_A \mid s'_i \nu'_i) & (s_d 2 \nu'_d q \mid s_d \nu_d) \\
= (-)^{s'_i + \nu'_i} & \sqrt{\frac{2 s'_i + 1}{5}} \sum_j \sqrt{(2 s_d + 1)(2j + 1)} W(s'_i s_A 2 s_d; s_d j) \\
& \times (s'_i j \nu'_i \mu \mid 2 - q)(s_d s_A \nu_d \nu_A \mid j - \mu),
\end{aligned} \tag{8.9}
$$

which allows

$$
\sum_{\nu_A} (s_d s_A \nu_d \nu_A \mid s_i \nu_i)(s_d s_A \nu_d \nu_A \mid j - \mu) = \delta_{j s_i} \delta_{\mu - \nu_i}. \tag{8.10}
$$

Further, by the use of

$$
\begin{aligned}
Y_{K\kappa}(\hat{\boldsymbol{k}}_f) Y^*_{K\kappa'}(\hat{\boldsymbol{k}}_f) \\
= (-)^{\kappa'} \sum_L \frac{2K + 1}{\sqrt{4\pi(2L + 1)}} (KK00 \mid L0) \\
\times (KK\kappa - \kappa' \mid LM) Y_{LM}(\hat{\boldsymbol{k}}_f),
\end{aligned} \tag{8.11}
$$

we finally get

$$N_R T_{2q}(K) = \sum_{s_i s_i' s_f} \sqrt{\frac{6}{5}(2s_i + 1)(2s_i' + 1)(2K + 1)}$$

$$\times (KK00 \mid 20)(-)^{s_i' - s_f}$$

$$\times W(s_i' s_A 2 s_d; s_d s_i) W(K s_i' K s_i; s_f 2)$$

$$\times F(s_i s_f K0) F(s_i' s_f K0)^* P_2^q(\cos\theta), \qquad (8.12)$$

where

$$N_R = \sum_{s_i s_f} |F(s_i s_f K0)|^2. \qquad (8.13)$$

The above results show that the cross section is isotropic and the tensor analyzing powers are proportional to $P_2^q(\cos\theta)$.

Because of the low incident energy, the dominant contribution to the reaction will arise from low partial waves in the final state. Since $l_f = K$, we will adopt only $K = 1$, neglecting $K = 3$, because higher-rank tensor interactions produce smaller contributions. Specifying $s_d = s_A = 1$ and $K = 1$ and using the numerical values for the Racah coefficients, we get T_{2q} from Eq. (8.12) as

$$T_{2q}(K = 1) = -\frac{2\sqrt{3}}{\sqrt{5}} \frac{1}{N_R} P_2^q(\cos\theta)$$

$$\times \left[\frac{1}{4\sqrt{3}}|F(11)|^2 - \frac{1}{20\sqrt{3}}|F(12)|^2 \right.$$

$$+ \frac{7}{20\sqrt{3}}|F(21)|^2 - \frac{7}{20\sqrt{3}}|F(22)|^2$$

$$+ \mathrm{Re}\left\{ \frac{2}{\sqrt{15}}F^*(01)F(21) - \frac{3}{2\sqrt{5}}F^*(11)F(21) \right.$$

$$\left. \left. + \frac{3}{10}F^*(12)F(22) \right\} \right], \qquad (8.14)$$

with

$$N_R = |F(01)|^2 + |F(02)|^2 + |F(11)|^2 + |F(12)|^2 + |F(21)|^2 + |F(22)|^2, \tag{8.15}$$

where $F(s_i s_f 10)$ is abbreviated as $F(s_i s_f)$. Equations (8.14) and (8.15) agree with those obtained by the general formula of analyzing powers in low-energy reactions [33]. In the following, for convenience, we will express Eq. (8.14) by

$$T_{2q} = -\frac{2\sqrt{3}}{\sqrt{5}} \alpha P_2^q (\cos\theta) \tag{8.16}$$

and treat α as a parameter, which is the object of further considerations.

8.3   Determination of the Spin Parity of Resonance

Since the incident wave is limited to the S-wave, the total angular momentum in the incident channel is equal to the channel spin s_i. Accordingly, when the system is in a resonance state, the resonance spin I_R is given by

$$I_R = s_i \tag{8.17}$$

and the parity is even because of the S-wave. In the present case, $s_i = 0, 1, 2$, for which we will examine the tensor analyzing powers.

(i) Assumption: $I_R = 0$.
   Neglecting $F(s_i s_f)$ for $s_i \neq 0$ in Eq. (8.14) gives

$$\alpha = 0, \quad T_{2q} = 0, \tag{8.18}$$

which cannot explain the experimental data.

(ii) Assumption: $I_R = 1$.
   Neglecting $F(s_i s_f)$ for $s_i \neq 1$ gives

$$\alpha = \frac{5 - p}{20\sqrt{3}(1 + p)}, \tag{8.19}$$

with

$$p \equiv \frac{|F(12)|^2}{|F(11)|^2}. \tag{8.20}$$

Since $0 \leq p \leq \infty$, the possible value of α is restricted to be

$$-\frac{1}{20\sqrt{3}} \leq \alpha \leq \frac{1}{4\sqrt{3}}. \tag{8.21}$$

(iii) Assumption: $I_R = 2$.

Neglecting $F(s_i s_f)$ for $s_i \neq 2$ in Eq. (8.14), we get

$$\alpha = \frac{7(p-1)}{20\sqrt{3}(1+p)}, \tag{8.22}$$

with

$$p = \frac{|F(21)|^2}{|F(22)|^2}. \tag{8.23}$$

Since $0 \leq p \leq \infty$, the possible value of α is restricted to be

$$-\frac{7}{20\sqrt{3}} \leq \alpha \leq \frac{7}{20\sqrt{3}}. \tag{8.24}$$

In Figs. 8.1(a) and 8.1(b), the possible T_{2q} ($q = 0, 1, 2$) within the limitations of α, Eqs. (8.21) and (8.24), are shown by the shaded area, together with the experimental data [34]. The calculations for the positive α are not shown because they do not reproduce the data. In the case shown in Fig. 8.1(a), where $I_R = 1$, the magnitudes of the calculated T_{2q} are too small when compared with those measured, while in the case seen in Fig. 8.1(b), where it has been assumed that $I_R = 2$, the shaded area covers most of the experimental data. This determines the resonance spin to be $I_R = 2$. Finally, by the use of Eq. (8.16), we can determine α so as to fit the experimental data. We get

$$\alpha = -0.167 \pm 0.018, \tag{8.25}$$

which satisfies the condition (8.24). The analyzing powers calculated with $\alpha = -0.167$ are shown by the solid lines, which reproduce the experimental data satisfactorily.

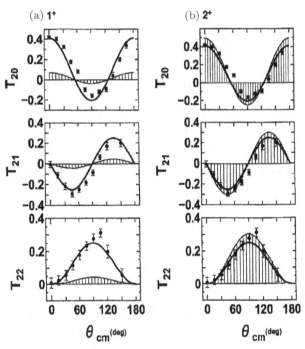

Fig. 8.1   Comparison of the tensor analyzing powers T_{2q} between the calculation by Eq. (8.16) and the experimental data in ^{6}Li$(d,p)^7$Li$(g.s.)$ reactions at $E_d = 90\,\mathrm{keV}$. (a) and (b) are for the assumptions, $I_R = 1$ and 2, respectively. The shaded areas show the possible T_{2q} ($q = 0, 1, 2$) calculated within the theoretical limits of α. The solid lines show the theoretical T_{2q} for $\alpha = -0.167$.

8.4   Reaction Amplitude in DWBA

In the preceding sections, the investigation has been performed by the invariant amplitude method, which is model-independent and quite general. However, the magnitudes of the analyzing powers are not predicted by this calculation and have been obtained by adjusting the parameter α so as to reproduce the measured analyzing powers. To derive theoretically the analyzing powers, which includes the magnitude, we need some reaction models. In the following, we will assume the stripping model for the reaction, and treat it using the distorted-wave Born approximation (DWBA) with the zero-range interaction. The calculated analyzing powers [35] are then compared with those obtained by the invariant amplitude method, as well as with the experimental data.

The DWBA amplitude for a stripping reaction $A(d,p)B$, $\langle f|M|i\rangle_{\mathrm{DWBA}}$, is given by

$$\langle f|M|i\rangle_{\mathrm{DWBA}} = \langle \Psi_f^{(-)}|v_{np}|\Psi_i^{(+)}\rangle, \tag{8.26}$$

where $\Psi_i^{(+)}(\Psi_f^{(-)})$ is the distorted wave of the $d-A$ $(p-B)$ relative motion, and is calculated when the $d-A$ $(p-B)$ potential is provided. Later, we will restrict the $d-A$ wave to the S-wave as in the preceding sections. The interaction v_{np} is chosen by considering the inverse process which is interpreted as the pickup of the neutron by the proton.

Introduce the channel spins, $s_i = s_d + s_A$ and $s_f = s_p + s_B$, and their spin functions $\chi_{s_i\nu_i}$ and $\chi_{s_f\nu_f}$. For example, for the incident channel

$$\chi_{s_d\nu_d}\chi_{s_A\nu_A} = \sum_{s_i\nu_i}(s_d s_A \nu_d \nu_A \mid s_i\nu_i)\chi_{s_i\nu_i}. \tag{8.27}$$

Using the channel spins, one can write $\Psi_i^{(+)}(\Psi_f^{(-)})$ by the distorted wave for each channel $\psi_{s_i\nu_i}^{(+)}(\psi_{s_f\nu_f}^{(-)})$ as

$$\Psi_i^{(+)} = \sum_{s_i\nu_i}(s_d s_A \nu_d \nu_A \mid s_i\nu_i)\psi_{s_i\nu_i}^{(+)} \tag{8.28}$$

and

$$\Psi_f^{(-)} = \sum_{s_f\nu_f}(s_p s_B \nu_p \nu_B \mid s_f\nu_f)\psi_{s_f\nu_f}^{(-)}, \tag{8.29}$$

which together lead to

$$\langle \Psi_f^{(-)}|v_{np}|\Psi_i^{(+)}\rangle$$
$$= \sum_{s_i\nu_i}\sum_{s_f\nu_f}(s_d s_A \nu_d \nu_A \mid s_i\nu_i)(s_p s_B \nu_p \nu_B \mid s_f\nu_f)$$
$$\times \langle \psi_{s_f\nu_f}^{(-)}|v_{np}|\psi_{s_i\nu_i}^{(+)}\rangle. \tag{8.30}$$

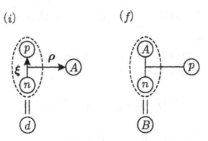

Fig. 8.2 Stripping model for $A(d,p)B$ reaction. (i) and (f) indicate the initial and final states of the reaction.

As is shown in Fig. 8.2, we will represent the relative coordinate for the $p - n$ motion by $\boldsymbol{\xi}$ and that for the $d - A$ motion by $\boldsymbol{\rho}$. The zero-range assumption for v_{np} gives

$$v_{np}\phi_d(\boldsymbol{\xi}) = D\delta(\boldsymbol{\xi}), \tag{8.31}$$

where $\phi_d(\boldsymbol{\xi})$ is the internal wave function of the deuteron and D is a constant properly determined. Describing the radial part of $\psi_{s_i\nu_i}^{(+)}$ by $R_{s_i}(k_i\rho)$, we get

$$v_{np}\psi_{s_i\nu_i}^{(+)} = \frac{D}{\sqrt{4\pi}}\delta(\boldsymbol{\xi})R_{s_i}(k_i\rho)\chi_{s_i\nu_i}. \tag{8.32}$$

To see the pure resonance effect on the analyzing power, we will ignore the contributions of spin-dependent interactions on the distorted wave. Therefore, we will only use the central interaction for the distortion of the outgoing wave in the final state.

$$\psi_{s_f\nu_f}^{(-)} = 4\pi \sum_{l_p m_p} (i)^{l_p} Y_{l_p m_p}^*(\hat{k}_f) Y_{l_p m_p}(\hat{\rho}) R_{l_p}\left(\frac{A}{A+1}k_f\rho\right)\chi_{s_f\nu_f},$$

$$\tag{8.33}$$

where $R_{l_p}(\frac{A}{A+1}k_f\rho)$ is the radial part of the final state wave function and $\chi_{s_f\nu_f}$ is given by

$$\chi_{s_f\nu_f} = \sum_{\nu_p\nu_B}(s_p s_B\nu_p\nu_B \mid s_f\nu_f)\chi_{s_p\nu_p}\chi_{s_B\nu_B}, \tag{8.34}$$

where $\chi_{s_B\nu_B}$ is the spin function of the final nucleus and is composed of the spin functions of the target nucleus $\chi_{s_A\nu_A}$ and the captured neutron $\Phi_{j_n\mu_n}$

$$\chi_{s_B\nu_B} = \sum_{\mu_n} (j_n s_A \mu_n \nu_A \mid s_B \nu_B) \Phi_{j_n\mu_n} \chi_{s_A\nu_A} \qquad (8.35)$$

with

$$\Phi_{j_n\mu_n} = \sum_{m_n} (l_n s_n m_n \nu_n \mid j_n \mu_n) \chi_{s_n\nu_n} Y_{l_n m_n}(\hat{\rho}) R_n(\beta\rho), \qquad (8.36)$$

where s_n, l_n and j_n are the spin, orbital angular momentum and total angular momentum of the captured neutron. Also, R_n is the radial part of the wave function of the neutron. For the convenience of further manipulations, we will change the order of the addition of the vectors by the use of the Racah coefficient, i.e. in the initial state

$$\text{from} \quad \boldsymbol{s}_p + \boldsymbol{s}_n = \boldsymbol{s}_d, \quad \boldsymbol{s}_d + \boldsymbol{s}_A = \boldsymbol{s}_i \quad \text{to} \quad \boldsymbol{s}_n + \boldsymbol{s}_A = \boldsymbol{J}, \quad \boldsymbol{J} + \boldsymbol{s}_p = \boldsymbol{s}_i$$
$$(8.37)$$

and in the final state

$$\text{from} \quad \boldsymbol{s}_n + \boldsymbol{l}_n = \boldsymbol{j}_n, \quad \boldsymbol{j}_n + \boldsymbol{s}_A = \boldsymbol{s}_B \quad \text{to} \quad \boldsymbol{s}_n + \boldsymbol{s}_A = \boldsymbol{J}', \quad \boldsymbol{J}' + \boldsymbol{l}_n = \boldsymbol{s}_B.$$
$$(8.38)$$

Because there is no spin-dependent interaction, $\boldsymbol{J}' = \boldsymbol{J}$, and we get

$$\langle \psi^{(-)}_{s_f\nu_f} | v_{np} | \psi^{(+)}_{s_i\nu_i} \rangle$$

$$= i^{l_n} \sqrt{4\pi(2s_d + 1)(2j_n + 1)} \sum_{J} (2J + 1) W(s_p s_n s_i s_A; s_d J)$$

$$\times W(l_n j_n J s_A; s_n s_B)$$

$$\times \sum_{M} (s_p J \nu_p M \mid s_i \nu_i) \sum_{m_n} (J l_n M m_n \mid s_B \nu_B)(s_p s_B \nu_p \nu_B \mid s_f \nu_f)$$

$$\times Y^*_{l_n m_n}(\hat{k}_f) I(s_i) \qquad (8.39)$$

with

$$I(s_i) = D \int R^*_{l_p = l_n} \left(\frac{A}{A+1} k_f \rho \right) R_n(\beta\rho) R_{s_i}(k_i\rho) \rho^2 d\rho. \qquad (8.40)$$

8.5 Tensor Analyzing Powers by DWBA

The tensor analyzing powers T_{2q} are given by Eq. (8.6), where the numerator of T_{2q} can be described in the channel spin representation as

$$\text{Tr}(M\tau_q^2 M^\dagger)$$
$$= \sum_{s_f \nu_f} \sum_{s_i \nu_i} \sum_{s_i' \nu_i'} \langle s_f \nu_f | M | s_i' \nu_i' \rangle \langle s_i' \nu_i' | \tau_q^2 | s_i \nu_i \rangle \langle s_f \nu_f | M | s_i \nu_i \rangle^*$$

$$(8.41)$$

and the Wigner–Eckart theorem gives

$$\langle s_i' \nu_i' | \tau_q^2 | s_i \nu_i \rangle = \frac{1}{\sqrt{2s_i'+1}} (s_i 2 \nu_i q \mid s_i' \nu_i')(s_i' \| \tau^2 \| s_i).$$

$$(8.42)$$

Using the DWBA amplitude Eq. (8.39) for $\langle s_f \nu_f | M | s_i \nu_i \rangle$ and Eq. (8.42) for the matrix element of τ_q^2, we get

$$\text{Tr}(M\tau_q^2 M^\dagger)$$

$$= \sum_{s_i s_i'} \frac{3\sqrt{2}}{5} (2s_B + 1)(2l_n + 1)(2j_n + 1)$$

$$\times \sqrt{(2s_i + 1)(2s_i' + 1)} (l_n l_n 00 \mid 20)$$

$$\times \sum_{JJ'} (2J + 1)(2J' + 1)(-)^{s_B - J'} W\left(\frac{1}{2}\frac{1}{2} s_i s_A; 1J\right)$$

$$\times W\left(\frac{1}{2}\frac{1}{2} s_i' s_A; 1J'\right) W\left(l_n j_n J s_A; \frac{1}{2} s_B\right) W\left(l_n j_n J' s_A; \frac{1}{2} s_B\right)$$

$$\times W\left(\frac{1}{2} J s_i' 2; s_i J'\right) W(J l_n J' l_n; s_B 2)(s_i' \| \tau^2 \| s_i)$$

$$\times \text{Re}\{I^*(s_i)I(s_i')\} P_2^q(\cos\theta)$$

$$(8.43)$$

with

$$(s_i' \| \tau^2 \| s_i) = (-)^{s_i'+1-s_A} \sqrt{15(2s_i + 1)(2s_i' + 1)} W(1s_i'1s_i; s_A 2).$$

$$(8.44)$$

In the derivation of (8.43), we used some transformations similar to Eq. (8.9). Also, Eq. (8.44) has been obtained by the prescription given in Ref. [3] with Eq. (2.7). The denominator of the analyzing powers is obtained in a similar way,

$$Tr(MM^\dagger) = \sum_{s_i} 3(2s_B \mid 1)(2j_n + 1)(2s_i + 1)$$

$$\times \sum_J (2J + 1)W^2 \left(\frac{1}{2}\frac{1}{2}s_i s_A; 1J \right)$$

$$\times W^2 \left(l_n j_n J s_A; \frac{1}{2} s_B \right) |I(s_i)|^2.$$

$$(8.45)$$

The tensor analyzing powers T_{2q} ($q = 0, 1, 2$),

$$T_{2q} = \frac{1}{Tr(MM^\dagger)} Tr(M\tau_q^2 M^\dagger)$$

$$(8.46)$$

can be calculated by the combination of Eqs. (8.43) and (8.45), which predicts the angular distribution of T_{2q} to be expressed by $P_2^q(\cos\theta)$. This qualitative nature of T_{2q} is consistent with the results of the preceding section, Eq. (8.16).

Now we will consider the $^6\text{Li}(d, p)$ $^7\text{Li}(g.s., \frac{3}{2}^-)$ reaction, assuming the reaction to be pure resonance, with the spin I_R. For this case, only one s_i is effective for the reaction because $s_i = I_R$. Also, the possible value of s_i is $s_i = 0, 1$ or 2, due to $s_i = s_{Li} + s_d$, which gives the candidate of I_R. For each I_R, we obtain T_{2q} for the reaction from Eq. (8.40) with Eqs. (8.43) and (8.45) by the use of the numericals of the CG coefficients and the Racah coefficients.

For

$$I_R = 0, \quad T_{2q} = 0,$$

$$(8.47)$$

for $\qquad I_R = 1 \quad j_n = \dfrac{1}{2}, \quad T_{2q} = -\dfrac{2}{7\sqrt{5}} P_2^q(\cos\theta),$ (8.48)

$$j_n = \frac{3}{2}, \quad T_{2q} = \frac{1}{10\sqrt{5}} P_2^q(\cos\theta),$$ (8.49)

and

for $\qquad I_R = 2 \quad j_n = \dfrac{1}{2}, \dfrac{3}{2}, \quad T_{2q} = \dfrac{14}{25\sqrt{5}} P_2^q(\cos\theta).$ (8.50)

These results are compared with the experimental data in Fig. 8.3 for the $^6\mathrm{Li}(d,p)\,^7\mathrm{Li}(\tfrac{3}{2}^-)$ reaction at $E_d = 90$ keV. In the figure it can be clearly seen that the data for $q = 0$, 1 and 2 are reproduced

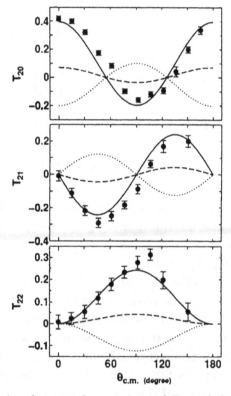

Fig. 8.3 Comparison between the experimental T_{2q} and the T_{2q} calculated by the DWBA. The dotted, dashed and solid lines are for the assumptions $I_R = 1$; $j_n = \tfrac{1}{2}$, $I_R = 1$; $j_n = \tfrac{3}{2}$ and $I_R = 2$; $j_n = \tfrac{1}{2}$ or $\tfrac{3}{2}$.

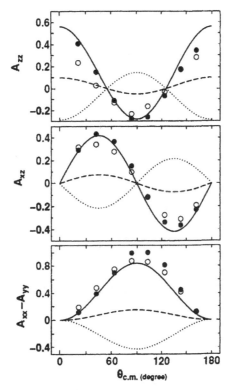

Fig. 8.4 Comparison between the experimental data of A_{zz}, A_{xz} and $A_{xx} - A_{yy}$ and the data calculated by the DWBA. For the definitions of the lines, see the caption of Fig. 8.3.

by the calculated T_{2q} for the value of $I_R = 2$, but not for the values of $I_R = 0$ or 1, and thus the resonance spin is determined to be 2. Further, present analyzing powers can be described by using α as

$$T_{2q} = -\frac{2\sqrt{3}}{\sqrt{5}}\alpha P_2^q(\cos\theta),$$

for which Eq. (8.50) gives

$$\alpha = -0.162. \tag{8.51}$$

This is consistent with the results obtained using the invariant amplitude method, that is, $\alpha = -0.167 \pm 0.018$.

Contributions of non-resonance reactions are investigated by including P-waves in the incident channel. The magnitude of such corrections is small, but slightly improves the agreement with the data of T_{2q} [34].

Finally, comments will be made on the analyzing powers of the same reaction at higher energies, $E_d = 600$ keV and 960 keV. The measurements [36] have been performed for the analyzing powers in the Cartesian coordinate system as A_{zz}, A_{xz}, $A_{xx} - A_{yy}$. We will transform the DWBA formulae, Eqs. (8.48)–(8.50) for T_{2q} to the A_{ij} scheme by using

$$A_{zz} = \sqrt{2}\,T_{20}, \tag{8.52}$$

$$A_{xz} = -\sqrt{3}T_{21}, \tag{8.53}$$

$$A_{xx} - A_{yy} = 2\sqrt{3}T_{22}, \tag{8.54}$$

where T_{2q} are given by Eqs. (8.48)–(8.50). Assuming the resonance reaction for these energies, we will compare the DWBA analyzing powers for the resonances with the experimental data in Fig. 8.4. The assumed value of $I_R = 2$ reproduces the data at both energies, 600 keV (closed circles) and 960 keV (open circles). This suggests that the reaction is induced through resonance with $I_R = 2$ at these energies.

Chapter 9

Depolarization in $p + {}^3$He Elastic Scattering and Time Reversal Theorem

Elastic scattering of protons by ^{3}He nuclei has attracted attention for many years. This is because in ^{3}He, two protons are coupled to each other predominantly in a spin-singlet state. Then $p - {}^3$He spin-dependent interactions, which include the ^{3}He spin, arise from interactions of the ^{3}He neutron with the incident proton. Thus $p + {}^3$He scattering is an important information source of the p–n spin-dependent interaction, since free neutrons are not useful for such investigations at present. In this chapter, we will examine the depolarization of the proton in the $p + {}^3$He elastic scattering, referring to experimental data at $E_p = 800$ MeV, with particular interest being focused on the spin-dependent interaction.

9.1 Characteristics of Observed Depolarization of Protons in Scattering by ^{3}He

The depolarization of a nucleon has been introduced in Chapter 3, which describes the polarization of the nucleon scattered, by a nucleus for instance, when the incident nucleon is polarized. Equation (3.46) gives the depolarization $D_i^j(\theta)$ for the proton, where

$$D_i^j(\theta) = \frac{1}{N_R} Tr(M\sigma_i M^\dagger \sigma_j) \tag{9.1}$$

with

$$N_R = Tr(MM^\dagger) = 4\sigma(\theta), \qquad (9.2)$$

where M is the T-matrix of the scattering, σ is the Pauli spin operator of the proton, and the suffixes i and j denote the direction of the polarization of the proton in the initial and final states, and takes either x, y or z in the Cartesian coordinate system. Such depolarizations, $D_x^x, D_z^z, D_x^z, D_z^x, D_y^y$, have been measured for the $p + {}^3$He elastic scattering at $E_p = 800$ MeV [37], which are shown in Fig. 9.1. As can be seen in this figure, the measured D_i^j satisfies the following

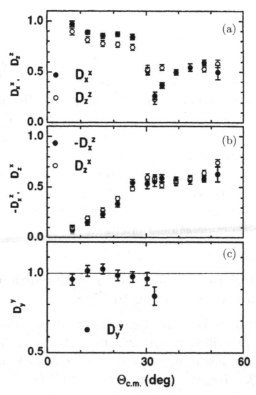

Fig. 9.1 Depolarizations of the $p + {}^3$He elastic scattering at $E_p = 800$ MeV. In (a) the solid and open circles are for D_x^x and D_z^z respectively. In (b) the solid and open circles are for $-D_x^z$ and D_z^x respectively. In (c) the solid circles are for D_y^y.

approximate relations,

$$D_x^x(\theta) \approx D_z^z(\theta), \quad D_z^x(\theta) \approx -D_x^z, \quad D_y^y(\theta) \approx 1, \qquad (9.3)$$

where the coordinate axes are chosen as $z \parallel \boldsymbol{k}_i, y \parallel \boldsymbol{k}_i \times \boldsymbol{k}_f$. The above results are quite interesting, because they suggest that some correction terms will fix the relations to the complete ones. In fact, in later sections, we will derive such complete relations, which are reduced to

$$D_x^x(\theta) = D_z^z(\theta), \quad D_z^x(\theta) = -D_x^z(\theta), \quad D_y^y(\theta) = 1, \qquad (9.4)$$

when the spin of the proton or that of ³He is ignored. That is, the correction term describes the contributions of the spin-dependent parts of the $p-$³He interaction, and is important to obtain the entire picture of the interaction.

9.2 Scattering Amplitude for Collision between Spin $\frac{1}{2}$ Particles

Because our interests lie in the study of the spin-dependent interactions, it is convenient to treat the scattering by the invariant amplitude method. According to the prescription given in Chapter 4, we can expand the T-matrix M into spin-space tensors S_κ^K,

$$M = \sum_{K\kappa} (-)^\kappa S_{-\kappa}^K R_\kappa^K, \qquad (9.5)$$

where $K(\kappa)$ is the rank (z-component) of the tensor and R_κ^K is the coordinate space tensor. Introducing the channel spins s_i and s_f and evaluating the matrix element of $S_{-\kappa}^K$ by the Wigner–Eckart theorem, we get

$$\langle s_f \nu_f; \boldsymbol{k}_f | M | s_i \nu_i; \boldsymbol{k}_i \rangle$$
$$= \sum_{K=0} (-)^{s_f - \nu_f} (s_i s_f \nu_i - \nu_f \mid K\kappa) M_\kappa^{(K)}(s_i s_f; \boldsymbol{k}_i, \boldsymbol{k}_f), \qquad (9.6)$$

where

$$M_\kappa^{(K)}(s_i s_f; \mathbf{k}_i, \mathbf{k}_f) = \frac{(-)^{s_i - s_f}}{\sqrt{2K+1}}(s_f \| S^K \| s_i)\langle \mathbf{k}_f | R_\kappa^K(s_i, s_f) | \mathbf{k}_i \rangle.$$

$$(9.7)$$

Here $\mathbf{K} = \mathbf{s}_i + \mathbf{s}_f$, where $s_i = 0, 1$, and $s_f = 0, 1$ because of $s_p = s_{\text{He}} = \frac{1}{2}$. Then the allowed K is either $K = 0, 1$ or 2. In Eq. (9.7), the terms $K = 0, 1$, and 2 express the contributions of the scalar, vector and tensor interactions, respectively. Further, as was discussed in Chap. 4, parity conservation gives

$$M_{-\kappa}^{(K)}(s_i s_f; \mathbf{k}_i, \mathbf{k}_f) = (-)^{K-\kappa} M_\kappa^{(K)}(s_i s_f; \mathbf{k}_i, \mathbf{k}_f) \qquad (9.8)$$

and the time reversal theorem gives, for the vector amplitude,

$$M_\kappa^{(1)}(10; \mathbf{k}_i, \mathbf{k}_f) = -M_\kappa^{(1)}(01; \mathbf{k}_i, \mathbf{k}_f), \qquad (9.9)$$

and for the tensor amplitudes,

$$\sqrt{\frac{3}{2}} M_0^{(2)}(11; \mathbf{k}_i, \mathbf{k}_f) - M_2^{(2)}(11; \mathbf{k}_i, \mathbf{k}_f) = -2 \cot \theta M_1^{(2)}(11; \mathbf{k}_i, \mathbf{k}_f).$$

$$(9.10)$$

9.3 Interaction Model for $p + {}^3\text{He}$ System

For further developments, it is convenient to refer to a practical model for the $p - {}^3\text{He}$ interaction. In this situation, we can adopt the following model proposed as a general form [38], which is given in the form of the scattering amplitude,

$$\langle \mathbf{k}_f | M | \mathbf{k}_i \rangle = U_\alpha + (\mathbf{s}_p \cdot \mathbf{s}_h) U_\beta + (\mathbf{s}_p \cdot \mathbf{n}) S_p + (\mathbf{s}_h \cdot \mathbf{n}) S_h$$
$$+ \hat{S}_T(\mathbf{n}) T_n + (\hat{S}_T(\mathbf{l}) - \hat{S}_T(\mathbf{m})) T_{lm}, \qquad (9.11)$$

where $\mathbf{s}_p$ and $\mathbf{s}_h$ are the spin operators of the proton and ${}^3\text{He}$, and $\hat{S}_T(\mathbf{a})$ is a tensor operator constructed by a unit vector $\mathbf{a}$ as

$$\hat{S}_T(\mathbf{a}) = 4\{3(\mathbf{s}_p \cdot \mathbf{a})(\mathbf{s}_h \cdot \mathbf{a}) - (\mathbf{s}_p \cdot \mathbf{s}_h)\}. \qquad (9.12)$$

Here $\boldsymbol{a}$ is can be $\boldsymbol{l}, \boldsymbol{m}$, or $\boldsymbol{n}$, where $\boldsymbol{l}$ and $\boldsymbol{n}$ are the unit vectors in the direction of $\boldsymbol{k}_i + \boldsymbol{k}_f$ and $\boldsymbol{k}_i \times \boldsymbol{k}_f$, respectively, and $\boldsymbol{m} = \boldsymbol{n} \times \boldsymbol{l}$. Then

$$\hat{S}_T(\boldsymbol{l}) + \hat{S}_T(\boldsymbol{m}) + \hat{S}_T(\boldsymbol{n}) = 0. \tag{9.13}$$

Here $U_\alpha, U_\beta, \ldots$ are the scattering amplitudes, which are functions of the scattering angle θ, and U_α and U_β stand for the spin-independent and spin–spin central forces, S_p and S_h for the LS interactions of the proton and ^{3}He. T_n and T_{lm} describe the contribution of the tensor interactions. These are related to $M_\kappa^{(K)}(s_i s_f; \boldsymbol{k}_i, \boldsymbol{k}_f)$ in the previous section as

$$M_0^{(0)}(00; \boldsymbol{k}_i, \boldsymbol{k}_f) = U_\alpha - \frac{3}{4}U_\beta, \tag{9.14}$$

$$M_0^{(0)}(11; \boldsymbol{k}_i, \boldsymbol{k}_f) = \sqrt{3}U_\alpha + \frac{\sqrt{3}}{4}U_\beta, \tag{9.15}$$

$$M_1^{(1)}(01; \boldsymbol{k}_i, \boldsymbol{k}_f) = \frac{i}{2\sqrt{2}}(S_p - S_h), \tag{9.16}$$

$$M_1^{(1)}(11; \boldsymbol{k}_i, \boldsymbol{k}_f) = -\frac{i}{2}(S_p + S_h), \tag{9.17}$$

$$M_0^{(2)}(11; \boldsymbol{k}_i, \boldsymbol{k}_f) = 3\sqrt{6}\cos\theta \, T_{lm} - \sqrt{6}T_n, \tag{9.18}$$

$$M_1^{(2)}(11; \boldsymbol{k}_i, \boldsymbol{k}_f) = -6\sin\theta \, T_{lm}, \tag{9.19}$$

$$M_2^{(2)}(11; \boldsymbol{k}_i, \boldsymbol{k}_f) = -3\cos\theta \, T_{lm} - 3T_n. \tag{9.20}$$

Due to Eq. (9.10), among the three tensor amplitudes, two are independent. Here, we can choose, as follows, T_α and T_β, as the two independent amplitudes,

$$T_\alpha \equiv \frac{1}{\sqrt{6}}M_0^{(2)}(11; \boldsymbol{k}_i, \boldsymbol{k}_f) + M_2^{(2)}(11; \boldsymbol{k}_i, \boldsymbol{k}_f) = -4T_n, \tag{9.21}$$

$$T_\beta \equiv -M_1^{(2)}(11; \boldsymbol{k}_i, \boldsymbol{k}_f)/\sin\theta = 6T_{lm}. \tag{9.22}$$

9.4 Relationships between Proton Depolarizations in $p + {}^3$He Scattering

We can describe the depolarizations of the proton in terms of the new scattering amplitudes $U_\alpha, U_\beta, \ldots, T_\beta$ to clarify the role of the proton-^{3}He spin-dependent interactions. Inserting Eqs. (9.14)–(9.20) into Eqs. (9.1) and (9.2) through Eq. (9.6), we get

$$D_x^x(\theta) = \frac{1}{\sigma(\theta)} \left[D(1) - \frac{1}{2} \cos\theta \, \mathrm{Re}\{(U_\beta + T_\alpha)^* T_\beta\} \right], \qquad (9.23)$$

and

$$D_z^z(\theta) = \frac{1}{\sigma(\theta)} \left[D(1) + \frac{1}{2} \cos\theta \, \mathrm{Re}\{(U_\beta + T_\alpha) T_\beta\} \right] \qquad (9.24)$$

with

$$D(1) = |U_\alpha|^2 - \frac{1}{16}|U_\beta|^2 + \frac{1}{4}(|S_h|^2 - |S_p|^2) - \frac{1}{4}|T_\alpha|^2 + \frac{1}{4}\mathrm{Re}\{U_\beta^* T_\alpha\} \qquad (9.25)$$

and

$$\sigma(\theta) = |U_\alpha|^2 + \frac{1}{16}|U_\beta|^2 + \frac{1}{4}(|S_p|^2 + |S_h|^2) + \frac{3}{8}|T_\alpha|^2 + \frac{1}{2}|T_\beta|^2. \qquad (9.26)$$

When the contribution of $D(1)$ dominates in the numerator of Eqs. (9.23) and (9.24), one can write the approximate relation

$$D_x^x(\theta) \approx D_z^z(\theta), \qquad (9.27)$$

which agrees with Eq. (9.3) observed in the experimental data. This suggests a characteristic of the contribution of the scattering amplitudes; that is, U_α, which is spin-independent, gives the largest contribution, and other amplitudes $U_\beta, \ldots, T_\beta$, which are spin-dependent, give minor contributions to D_x^x and D_z^z. Since U_α is included in $D(1)$ but not in the second term of the numerator of Eqs. (9.23) and (9.24), the above speculation leads to Eq. (9.27). The exact relations between D_x^x and D_z^z are derived from Eqs. (9.23) and (9.24):

$$D_x^x(\theta) - D_z^z(\theta) = -\frac{\cos\theta}{\sigma(\theta)}\text{Re}\{(U_\beta + T_\alpha)T_\beta\} \qquad (9.28)$$

and

$$D_x^x(\theta) + D_z^z(\theta) = \frac{2}{\sigma(\theta)}D(1). \qquad (9.29)$$

The right-hand side of Eq. (9.28) provides the information of the spin-dependent amplitude, U_β, T_α and T_β in the form of $\text{Re}\{(U_\beta+T_\alpha)^*T_\beta\}$. This quantity is proportional to T_β and in this sense, $D_x^k - D_z^z$ is a measure of the magnitude of the tensor amplitude T_β. When the spin-dependent interactions are very weak, Eq. (9.29) gives an approximate relation $D_x^x + D_z^z \approx 2$. However, the present experimental data satisfy this relation only at $\theta \approx 8°$. Further applications of Eq. (9.29) will be given later.

The consideration developed above is based on the assumption that the magnitude of U_α is larger than those of other amplitudes. In the following, as an example, we evaluate the spin-independent central interaction V_0 and the spin–spin central interaction V_σ quantitatively. These interactions are the sources of the amplitudes U_α and U_β. The interactions V_0 and V_σ are obtained by the folding model where the nucleon–nucleon interaction [39] between the proton and the nucleon of ^{3}He are folded by using the ^{3}He internal wave function. The ^{3}He wave function is calculated by the Faddeev method, using the AV18 two-nucleon force and the Brazil model three-nucleon force [40]. More details will be referred to in the next chapter. The calculated values of V_0 and V_σ [41] are shown in Fig. 9.2, as a function of the $p - {}^3$He distance, R. As can be easily seen, V_σ is much smaller than V_0 for a small R. However, for a large R, for instance $R \gtrsim 1.5$ fm, the magnitude of $\text{Re}\{V_\sigma\}$ is comparable to that of V_0. Thus, the contribution of V_σ to the scattering amplitude will be appreciable at some angles.

Next, we will consider the relationship between $D_z^x(\theta)$ and $D_x^z(\theta)$, for which the experimental data predict

$$D_z^x(\theta) \approx -D_x^z(\theta). \qquad (9.30)$$

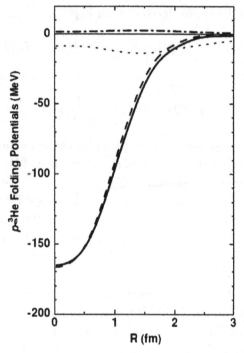

Fig. 9.2 The central interactions V_0 and V_σ between the proton and ^{3}He. The spin-independent central interaction V_0, and the spin–spin central interaction V_σ are shown by the solid (dashed) line and the dotted (dash-dotted) line for the real (imaginary) part respectively. R is the $p - {}^3$He distance.

Inserting Eq. (9.6) into Eqs (9.1) and utilizing Eqs. (9.14)–(9.20), we get

$$D_x^z(\theta) = \frac{1}{\sigma(\theta)} \left[-D(2) + \frac{1}{2} \sin\theta \; \text{Re}\{(U_\beta + T_\alpha)^* T_\beta\} \right] \qquad (9.31)$$

and

$$D_z^x(\theta) = \frac{1}{\sigma(\theta)} \left[D(2) + \frac{1}{2} \sin\theta \; \text{Re}\{(U_\beta + T_\alpha)^* T_\beta\} \right] \qquad (9.32)$$

with

$$D(2) = \text{Re} \left\{ i \left(U_\alpha^* S_p - \frac{1}{4} U_\beta^* S_h - \frac{1}{2} S_h^* T_\alpha \right) \right\}. \qquad (9.33)$$

Here, $D(2)$ dominates the numerators of Eqs. (9.30) and (9.32), because the magnitude of U_α is larger than those of other amplitudes, and $D(2)$ includes U_α, but the second term of the numerator does not. Then, as an approximation, we will take only $D(2)$ in the numerators of the equations concerned and get

$$D_x^z(\theta) = -D_z^x(\theta). \tag{9.34}$$

This result justifies the predicted outcome of the experiment. The full relationship between D_x^z and D_z^x is obtained by taking into account the second term of the numerators,

$$D_x^z(\theta) + D_z^x(\theta) = \frac{1}{\sigma(\theta)} \sin\theta \ \text{Re}\{((U_\beta + T_\alpha)^* T_\beta)\}. \tag{9.35}$$

The right-hand side of the above equation gives the information of $\text{Re}\{(U_\beta + T_\alpha)^* T_\beta\}$, which is similar to that given by $D_x^x - D_z^z$. However, $(D_x^x - D_z^z)$ has the factor $\cos\theta$, which decreases the magnitude of $\text{Re}\{U_\beta + T_\alpha^* T_\beta\}$, at angles around $\theta = 90°$. On the other hand, $D_x^z + D_z^x$, which has the factor $\sin\theta$ instead of $\cos\theta$, supplies the information of $\text{Re}\{(U_\beta + T_\alpha)^* T_\beta\}$ at such angles.

Finally, we can examine D_y^y by describing it in terms of $U_\alpha, \ldots, T_\beta$.

$$D_y^y(\theta) = 1 - \frac{1}{\sigma(\theta)} \left\{ \frac{1}{4} |U_\beta + T_\alpha|^2 + |T_\beta|^2 \right\}, \tag{9.36}$$

where $\sigma(\theta)$ is given by Eq. (9.26). In the second term of the right-hand side of the above equation, $\sigma(\theta)$ includes $|U_\alpha|^2$ but the numerator does not. Therefore, this term will be small compared to 1. Neglecting the term supports the prediction by the experimental data,

$$D_y^y(\theta) \approx 1. \tag{9.37}$$

Now, we can make a few remarks on the relationships between proton depolarizations. Equation (9.29), with the help of Eqs. (9.25) and (9.36) provides, as follows

$$D_x^x(\theta) + D_y^y(\theta) + D_z^z(\theta)$$

$$= 3 - \frac{1}{\sigma(\theta)} \left\{ \frac{3}{4}|U_\beta|^2 + \frac{3}{2}|T_\alpha|^2 + 2|T_\beta|^2 + |S_p|^2 \right\}. \qquad (9.38)$$

The right-hand side of the equation shows that $\sum_{i=x,y,z} D_i^i$ consists of the main part, 3, and a correction term which is the contribution of the spin-dependent interactions. The present experimental data show that this correction is not negligible even at small scattering angles.

From Eqs. (9.28) and (9.35), we get

$$\frac{D_x^z(\theta) + D_z^x(\theta)}{D_z^z(\theta) - D_x^x(\theta)} = \tan\theta, \qquad (9.39)$$

which is independent of the interaction model. The relation is similar to the one required for the p-n interaction as the time reversal invariance. The $p - {}^3\text{He}$ interaction must satisfy the relation in Eq. (9.39), as a test of the time reversal invariance.

Chapter 10

Three Nucleon Force and Polarization Phenomena in a Three Nucleon System

10.1 Three Nucleon Force

In the previous chapters, the nuclear interaction in reactions has been considered to be induced by the interaction between two nucleons, the main part of which is produced by the one-pion exchange mechanism, as in Fig. 10.1(a). However, for three nucleon systems, a new type of interaction has been derived, by taking into account the exchange of two pions between the three nucleons, as illustrated in Fig. 10.1(b) [42]. This suggests that it may be the additional source of the nuclear interaction in the reaction. In Fig. 10.1(b), the pion emitted by nucleon 1 is absorbed by nucleon 2, and the pion emitted by nucleon 2 is absorbed by nucleon 3. Since the absorption and subsequent emission of the pion by nucleon 2 is equivalent to the scattering of the pion by nucleon 2, the whole process has been treated by using the dispersion relation. During this process, nucleon 2 can be excited to the $I = T = \frac{3}{2}$ resonance state by the pion absorption. Due to this resonance, such virtual excitation enhances its contribution to physical quantities. For example, a correction to the binding energy of a triton on the order of 1 MeV has been obtained using this mechanism. In this case, the interaction shown by Fig. 10.1(b) has been named the nuclear three-body force (here abbreviated as the three-nucleon force or $3NF$ for simplicity) and has been treated through

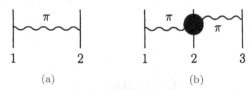

Fig. 10.1   Nuclear force by pion exchange. (a) Two nucleon force by one-pion exchange. (b) Three nucleon force by two-pion exchange. The closed circle expresses the $I = T = \frac{3}{2}$ resonance.

various approximations. At present, we will adopt the Brazil model for the $3NF$ [40], which will be hereafter abbreviated as $Br3NF$. In this chapter, we will investigate polarization phenomena in three-nucleon systems, for which the contribution of the three-nucleon force is examined, and compared with experimental data.

As is well known, the three-nucleon system cannot be solved in any analytic way, and several methods have been proposed to obtain numerical solutions, among which we will adopt the Faddeev method [43, 44]. As the input of the calculation, we will take the AV18 two-nucleon force [45] as the standard. The calculations are performed for the $p + d$ ($n + d$) elastic scattering at $E_p(E_n) = 3$ MeV, where the proton–proton Coulomb interaction is taken into account for the $p + d$ system.

In the following sections, the scattering amplitudes are analyzed by the invariant amplitude method, accompanied by numerical examinations. The cross section and the vector and tensor analyzing powers are calculated. These results are compared with experimental data.

10.2   Nucleon Deuteron Scattering Amplitudes by the Invariant-Amplitude Method

The scattering amplitude of $p + d$ elastic scattering is written in terms of the matrix element of the T-matrix M as $\langle \nu_{p_f}, \nu_{d_f}; \mathbf{k}_f | M | \nu_{p_i}, \nu_{d_i}; \mathbf{k}_i \rangle$, where ν is the z component of the spin s, for example $\nu_{p_i}(\nu_{p_f})$ is the z component of the proton spin $\mathbf{s}_p$ in the initial (final) state, and $\mathbf{k}_i(\mathbf{k}_f)$ is the $p - d$ relative momentum in the incident (outgoing) channel. Let us expand M into the tensor component $M^{(K)}$. The matrix element of $M^{(K)}$ is given by the invariant amplitude method. Equation (4.5) gives

$$\langle \nu_{p_f}, \nu_{d_f}; \boldsymbol{k}_f | M^{(K)} | \nu_{p_i}, \nu_{d_i}; \boldsymbol{k}_i \rangle$$

$$= \sum_{s_i s_f} (s_p s_d \nu_{p_i} \nu_{d_i} \mid s_i \nu_i)(s_p s_d \nu_{p_f} \nu_{d_f} \mid s_f \nu_f)(-)^{s_f - \nu_f}$$

$$\times (s_i s_f \nu_i - \nu_f \mid K\kappa)$$

$$\times \sum_{l_i = \bar{K} - K}^{K} [C_{l_i}(\hat{k}_i) \otimes C_{l_f = \bar{K} - l_i}(\hat{k}_f)]_\kappa^K F(s_i s_f K l_i), \qquad (10.1)$$

where $F(s_i s_f K l_i)$ is the invariant amplitude defined in Chap. 4, but the variable E and $\cos\theta$ are not described explicitly for simplicity.

In the above equation, s_i and s_f take the respective values of $\frac{1}{2}$ and $\frac{3}{2}$ due to $s_i(s_f) = s_p + s_d$. Since $K = s_i + s_f$, we will define the scalar $(K = 0)$ amplitudes U_i $(i = 1, 3)$, the vector $(K = 1)$ amplitudes S_i $(i = 1, 2, 3, 4)$, the tensor $(K = 2)$ amplitude $T_i(\kappa)$ $(i = 1, 2, 3; \kappa = 0, 1, 2)$ and the third-rank tensor $(K = 3)$ amplitudes $V(\kappa)$ $(\kappa = 1, 2, 3)$ as follows:

$$U_1 \equiv F\left(\frac{1}{2} \frac{1}{2} 00\right), \quad U_3 \equiv F\left(\frac{3}{2} \frac{3}{2} 00\right), \qquad (10.2)$$

$$S_1 \equiv [C_1(\hat{k}_i) \otimes C_1(\hat{k}_f)]_1^1 F\left(\frac{1}{2} \frac{1}{2} 11\right), \qquad (10.3)$$

$$S_2 \equiv [C_1(\hat{k}_i) \otimes C_1(\hat{k}_f)]_1^1 F\left(\frac{3}{2} \frac{1}{2} 11\right), \qquad (10.4)$$

$$S_3 \equiv [C_1(\hat{k}_i) \otimes C_1(\hat{k}_f)]_1^1 F\left(\frac{3}{2} \frac{3}{2} 11\right), \qquad (10.5)$$

$$S_4 = -S_2, \qquad (10.6)$$

$$T_1(\kappa) \equiv \sum_{l_i} [C_{l_i}(\hat{k}_i) \otimes C_{l_f}(\hat{k}_f)]_\kappa^2 F\left(\frac{3}{2} \frac{1}{2} 2l_i\right), \qquad (10.7)$$

$$T_2(\kappa) \equiv \sum_{l_i} [C_{l_i})(\hat{k}_i) \otimes C_{l_f}(\hat{k}_f)]_\kappa^2 F\left(\frac{1}{2}\ \frac{3}{2}\ 2l_i\right), \qquad (10.8)$$

$$T_3(\kappa) \equiv \sum_{l_i} [C_{l_i})(\hat{k}_i) \otimes C_{l_f}(\hat{k}_f)]_\kappa^2 F\left(\frac{3}{2}\ \frac{3}{2}\ 2l_i\right) \quad \kappa = 0, 1, 2,$$

$$(10.9)$$

$$V(\kappa) \equiv \sum_{l_i} [C_{l_i})(\hat{k}_i) \otimes C_{l_f}(\hat{k}_f)]_\kappa^3 F\left(\frac{3}{2}\ \frac{3}{2}\ 3l_i\right) \quad \kappa = 1, 2, 3.$$

$$(10.10)$$

To see the essential features of these amplitudes, we will perform numerical calculations, tentatively neglecting the Coulomb interaction. In Fig. 10.2, the $n + d$ scattering amplitudes at $E_n = 3$ MeV

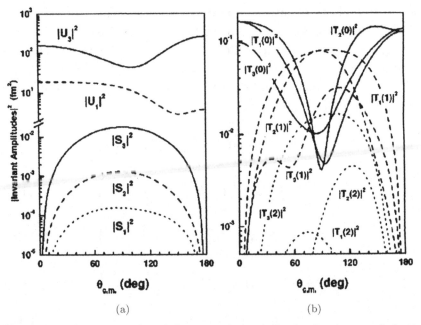

(a)   (b)

Fig. 10.2 The magnitudes of the scattering amplitudes for the $n + d$ elastic scattering at $E_n = 3$ MeV. $|U_1|^2$, $|U_3|^2$ are for the scalar amplitudes, $|S_1|^2$, $|S_2|^2$, $|S_3|^2$ are for the vector amplitudes and $|T_1(0)|^2$, $|T_2(0)|^2$, $|T_3(0)|^2$, $|T_1(1)|^2$, $|T_2(1)|^2$, $|T_3(1)|^2$, $|T_1(2)|^2$, $|T_2(2)|^2$, $|T_3(2)|^2$ are for the tensor amplitudes.

are displayed in the form of the absolute squares, i.e. $|U_i|^2$ ($i = 1, 3$), $|S_i|^2$ ($i = 1, 2, 3$) and $|T_i(\kappa)|^2$ ($i = 1, 2, 3; \kappa = 0, 1, 2$), as functions of the scattering angle θ [46]. The third-rank tensor amplitudes $V(\kappa)$ ($\kappa = 1$–3) are not given because their magnitudes are very small. The coordinate axes are chosen as $z \parallel \mathbf{k}_i$ and $y \parallel \mathbf{k}_i \times \mathbf{k}_f$. The input interaction is the AV18 two nucleon force, but the three nucleon force is not included. In the figure, the magnitudes of the scalar amplitudes, $|U_1|^2$ and $|U_3|^2$ are much larger than those of the other amplitudes, $|S_i|^2$ ($i = 1, 2, 3$) and $|T_i(\kappa)|^2$ ($i = 1$–$3; \kappa = 0$–2). In more detail, $|U_3|^2$ is larger than $|U_1|^2$ and $|S_3|^2 > |S_2|^2 > |S_1|^2$. Many tensor amplitudes compete with one another, particularly at angles around $\theta = 90°$. Since the scalar amplitudes are dominant over other amplitudes, for a qualitative examination of physical quantities, it will be a good approximation to consider only the terms which include the scalar amplitudes U_1 or U_3. We describe this approximation as the scalar amplitude dominance (SAD) approximation.

10.3 Observables and Comparison with Experimental Data

The observables obtained by numerical calculations [47] are compared with the experimental data [48] in the $p + d$ elastic scattering at $E_p = 3$ MeV. Typically, we treat the cross section $\frac{d\sigma}{d\Omega}$, the vector analyzing power of the proton A_y and the vector and tensor analyzing powers of the deuteron iT_{11} and $T_{2\kappa}$ ($\kappa = 0, 1, 2$). Their expressions are given in Chap. 3.

$$\frac{d\sigma}{d\Omega} = \frac{1}{6}N_R, \quad N_R = Tr(MM^\dagger), \tag{10.11}$$

$$A_y = \frac{1}{N_R} Tr(M\sigma_y M^\dagger), \tag{10.12}$$

$$iT_{11} = \frac{i}{N_R} Tr(M\tau_1^1 M^\dagger), \tag{10.13}$$

$$T_{2\kappa} = \frac{1}{N_R} Tr(M\tau_\kappa^2 M^\dagger) \quad \kappa = 0, 1, 2. \tag{10.14}$$

A comparison between the calculation and experimental data [47] is shown in Fig. 10.3 [46], where several combinations of the interactions are considered as the input of the calculation. Specifically, two kinds of $3NF$, V_{GS3NF} and V_{SO3NF} are newly introduced. The interaction $Br3NF$ has two components from the viewpoint of the tensorial character in the spin space, i.e. the scalar and the second-rank tensor. To distinguish the contribution of the scalar component from that of the tensor, we will introduce V_{GS3NF}, defined below as representing the scalar component.

$$V_{GS3NF} = V_0^G \sum_{i \neq j \neq k} \exp\left\{ - \left(\frac{r_{ij}}{r_G} \right)^2 - \left(\frac{r_{ki}}{r_G} \right)^2 \right\} \tag{10.15}$$

with

$$V_0^G = -40\,\text{MeV}, \quad r_G = 1.0\,\text{fm}. \tag{10.16}$$

The calculation by the $AV18+Br3NF$ interaction gives 7.79 MeV for the ^{3}He binding energy and that of the $AV18 + V_{GS3NF}$ interaction gives 7.74 MeV, both of which reproduce the empirical value 7.72 MeV to good approximation. Further, one can see that the scalar amplitudes U_1 and U_3 calculated by the V_{GS3NF} are very similar to those by the $Br3NF$ at $E_n = 3$ MeV. Then, V_{GS3NF} will be considered to simulate the scalar part of the $Br3NF$ to a good approximation.

As will be discussed later, the magnitudes of A_y and iT_{11} calculated with the $AV18$ interaction are remarkably smaller than those measured. To solve the discrepancies, an LS type three-nucleon force, V_{SO3NF} has been proposed phenomenologically [48]:

$$V_{SO3NF} = \frac{1}{2} W_0 \exp\{-\alpha\rho\} \sum_{i>j} [l_{ij}(\sigma_i + \sigma_j)]\hat{P}_{11},$$

$$\rho^2 = \frac{2}{3}(r_{12}^2 + r_{23}^2 + r_{31}^2), \tag{10.17}$$

where $\alpha = 1.5$ fm^{-1}, $W_0 = -20$ MeV and $\hat{P}_{11}$ is the projection operator of the $(1, 1)$ nucleon pair in the isospin triplet state. The contribution of V_{SO3NF} is examined in the numerical calculation.

In Fig. 10.3, the calculations by the interactions $AV18$, $AV18 + Br3NF$, $AV18 + V_{GS3NF}$, and $AV18 + Br3NF + V_{SO3NF}$ are shown by the solid, dashed, dotted and dash-dotted lines, respectively. These calculations reproduce the gross structures of the angular distribution of most observables. However, the magnitudes of the vector analyzing powers, A_y and iT_{11}, calculated by the interactions $AV18$, $AV18 + Br3NF$ and $AV18 + V_{G3NF}$ are much smaller than those measured, although the discrepancies vanish when V_{SO3NF} is included.

10.4 Contributions of the Three Nucleon Forces

Here, we will investigate the contribution of each interaction in more detail, with particular interest being focused on the three-nucleon forces. First, we will consider the characteristics of the observables using the SAD approximation, which gives

$$N_R = |U_1|^2 + |U_3|^2. \tag{10.18}$$

Accordingly, the cross section is described by the sum of the doublet $\sigma_1(\theta)$ and the quartet $\sigma_3(\theta)$ as

$$\frac{d\sigma}{d\Omega} = \sigma_1(\theta) + \sigma_3(\theta) \quad \text{with} \quad \sigma_1(\theta) = \frac{1}{6}|U_1|^2 \quad \text{and} \quad \sigma_3(\theta) = \frac{1}{6}|U_3|^2. \tag{10.19}$$

The vector analyzing powers are given by

$$A_y = \frac{4}{3N_R}\text{Im}\left\{ U_1\left(-\frac{1}{\sqrt{2}}S_1 + 2S_2\right)^* + U_3\left(\sqrt{2}S_2 + \sqrt{\frac{5}{2}}S_3\right)^* \right\}, \tag{10.20}$$

$$iT_{11} = \frac{1}{N_R}\sqrt{\frac{2}{3}}\text{Im}\{U_1(2S_1 - \sqrt{2}S_2)^* + U_3(-S_2 + \sqrt{5}S_3)^*\} \tag{10.21}$$

and the tensor analyzing powers are given by

$$T_{2\kappa} = \frac{1}{N_R}\mathrm{Re}\{-2U_1 T_1(\kappa)^* + \sqrt{2}U_3(T_2(\kappa) + T_3(\kappa))^*\}, \quad \kappa = 0, 1, 2.$$

(10.22)

Next, we will examine the effect of the $3NF$ interaction on the magnitudes of the scattering amplitudes included in the above

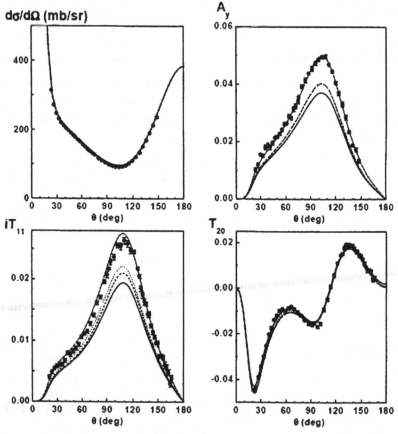

Fig. 10.3 The cross section $\frac{d\sigma}{d\Omega}$, the vector analyzing power of the proton A_y, the vector and tensor analyzing power of the deuteron, iT_{11} and $T_{2\kappa}$ ($\kappa = 0, 1, 2$) for the $p + d$ elastic scattering at $E_p = 3$ MeV. The solid, dashed, dotted and dot-dashed lines are the Faddeev calculations with the interactions $AV18$, $AV18 + Br3NF$, $AV18 + GS3NF$, $AV18 + Br3NF + V_{SO3NF}$, respectively. The experimental data are taken from Ref. [47].

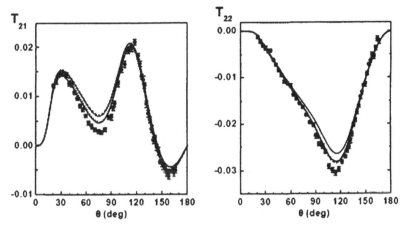

Fig. 10.3 (*Continued*)

equations. To avoid the Coulomb-interaction effect we will treat the $n+d$ scattering. In order to clarify the $3NF$ effects, we will display in Fig. 10.4 the scattering amplitudes by the $AV18$ two-nucleon force with the $3NF$ contributions in the form of the ratio to the $AV18$ scattering amplitudes without the $3NF$ contributions. There, the scalar amplitude U_1 is strongly affected by the $Br3NF$, as well as by the V_{GS3NF} interaction, making a sharp dip at large angles in the angular distribution. On the other hand, U_3 is hardly affected by such $3NF$ interactions. For the tensor amplitudes, we will consider $T_3(\kappa)$ with $\kappa = 0, 1, 2$ as the representative of the amplitudes. As is seen in (c) of the same figure, the calculated values of $T_3(0)$ and $T_3(1)$ are appreciably affected by the $Br3NF$, but $T_3(2)$ is not. The vector amplitudes S_1, S_2 and S_3 are enhanced uniformly by the V_{SO3NF}, except for S_1 at small angles. These effects yield characteristic behaviors in the angular distribution of the observables, as will be seen in the following.

Using Eq. (10.19) with the calculated U_1 and U_3, one can extract $\sigma_1(\theta)$ and $\sigma_3(\theta)$ from the full calculation, which is shown in Fig. 10.5 for the $n + d$ and $p + d$ scattering. As expected from the behavior of U_1 and U_3 in Fig. 10.4, the effect of the $3NF$ is remarkable in $\sigma_1(\theta)$ but not in $\sigma_3(\theta)$. The interactions of $Br3NF$ and V_{GS3NF} provide similar contributions to $\sigma_1(\theta)$, and these $3NF$ effects give

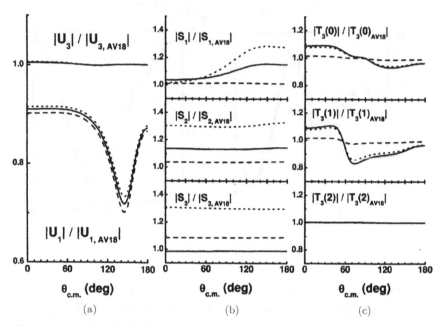

Fig. 10.4   The $3NF$ effects on the scattering amplitudes in the $n + d$ elastic scattering at $E_n = 3$ MeV. The magnitudes of the amplitudes with the $3NF$ are shown by the ratio to those without the $3NF$. The solid, dashed and dotted lines are for the calculations by the $AV18 + Br3NF$, $AV18 + V_{GS3NF}$ and $AV18 + Br3NF + V_{SO3NF}$ interactions, respectively.

good agreement with the experimental data of $\sigma_1(\theta)$ for the $p + d$ scattering, obtained from the measured cross section by subtracting the theoretical $\sigma_3(\theta)$.

As seen in the SAD approximation, the contributions of S_1, S_2 and S_3 are mixed up in the vector analyzing powers A_y and iT_{11}. However, $|S_1|^2$ is very small as evident in Fig. 10.2, and we will examine the two larger amplitudes S_2 and S_3. To separate the contribution of one from the other, we will consider the following linear combinations of the vector analyzing powers,

$$A_y - \frac{2}{\sqrt{3}}iT_{11} = \frac{2\sqrt{2}}{N_R}\mathrm{Im}\{-U_1 S_1^* + (\sqrt{2}U_1 + U_3)S_2^*\},$$

$$A_y + \frac{4}{\sqrt{3}}iT_{11} = \frac{2\sqrt{2}}{N_R}\mathrm{Im}\{U_1 S_1^* + \sqrt{5}U_3 S_3^*\}. \tag{10.23}$$

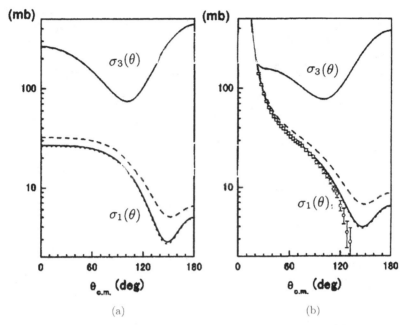

Fig. 10.5 The components of the cross section, $\sigma_1(\theta)$ and $\sigma_3(\theta)$ for the $n + d$ scattering (a) and for the $p + d$ scattering (b) at $E_n = E_p = 3$ MeV. The dashed, solid and dotted lines are for the interactions. $AV18$, $AV18 + Br3NF$ and $AV18 + V_{GS3NF}$, respectively.

In the above combinations, $A_y - \frac{2}{\sqrt{3}}iT_{11}$ clearly shows the contributions of S_2, and $A_y + \frac{4}{\sqrt{3}}iT_{11}$, that of S_3, because $|U_3| \gg |U_1|$. Comparisons with experimental data show that the interaction V_{SO3NF} accurately describes the contribution of the amplitude S_3, but not sufficiently that of the amplitude S_2.

Finally, we will investigate the contributions of the $3NF$ interaction to the tensor analyzing powers. Applying the prescription used for the scattering amplitudes, Fig. 10.7 shows the observables calculated with the $AV18 + 3NF$ interactions, as a ratio to those with only the $AV18$ interaction [49]. Experimental data are also expressed on the same scale, i.e. those divided by the $AV18$ calculation are plotted. Such representations are useful even for the examination of the cross section. As is seen in the figure, it is quite clear that the $3NF$ interactions improve the agreement between the calculated and

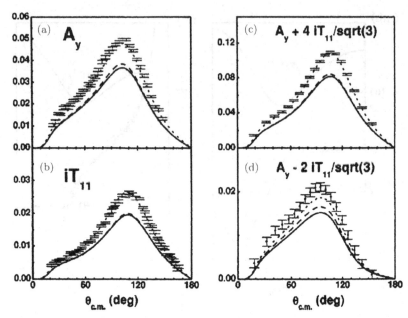

Fig. 10.6   The vector analyzing powers in the $p + d$ scattering at $E_p = 3$ MeV. The calculated $A_y, iT_{11}, A_y + \frac{4}{\sqrt{3}} iT_{11}$ and $A_y - \frac{2}{\sqrt{3}} iT_{11}$ are compared with the experimental data [46]. The solid, dashed and dotted lines are for the interactions $AV18$, $AV18 + Br3NF$ and $AV18 + Br3NF + V_{SO3NF}$.

measured cross sections. In the tensor analyzing powers, T_{20} and T_{21}, remarkable discrepancies are seen between the values calculated with the $3NF$ and those experimentally measured. For example, $\frac{T_{21}}{T_{21,AV18}}$ calculated with the $AV18 + 3NF$ at $\theta = 50° - 100°$ is situated on the opposite side of the experimental data with respect to the $2NF$ line $T_{21}/T_{21,AV18} = 1$. This suggests that the sign of the tensor part of the $3NF$ might be changed to fit the data. In the calculation of T_{22}, the contribution of the $Br3NF$ is very close to that of the V_{GS3NF} interaction. This means that the scalar part of the $Br3NF$ is important in T_{22}. Further, the calculated value of T_{22} is situated on the same side as the experimental data with respect to the $2NF$ line, $\frac{T_{22}}{T_{22,AV18}} = 1$, in reasonable agreement with the data.

For a quantitative examination of the above speculations on the sign of the tensor component of the $3NF$, we will first extend the simulation of the $Br3NF$ by Eq. (10.15), so as to include the tensor

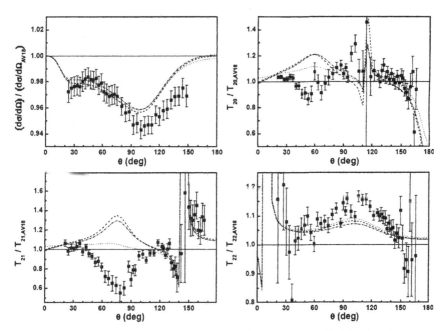

Fig. 10.7 The cross section and tensor analyzing powers divided by those calculated with the $AV18$ interaction for the $p + d$ scattering at $E_p = 3$ MeV. The dashed, dotted and dash-dotted curves are for the calculation with the interaction $AV18 + Br3NF$, $AV18 + V_{GS3NF}$ and $AV18 + Br3NF + V_{SO3NF}$, respectively.

interaction. That is, we will consider the following V as a substitute for the $Br3NF$.

$$V = \sum_{i \neq j \neq k} \exp\left\{ -\left(\frac{r_{ik}}{r_G}\right)^2 - \left(\frac{r_{jk}}{r_G}\right)^2 \right\} \left\{ V_0 + V_T S_T(ij)\hat{P}_{11} \right\},$$

(10.24)

where $\hat{P}_{11}$ is the same projection operator as the one in Eq. (10.17) and r_G is fixed to 1.0 fm. V_0 and V_T are the strengths of the scalar and tensor interactions. To reproduce the results of the $Br3NF$ calculation, we adopt $V_T = 20$ MeV and $V_0 = -38$ MeV, which give the empirical value of the ^{3}He binding energy, and produce cross sections similar to those obtained by V_{GS3NF}. As is shown in Fig. 10.8, the simulation is quite successful for T_{20}, T_{21} and T_{22}.

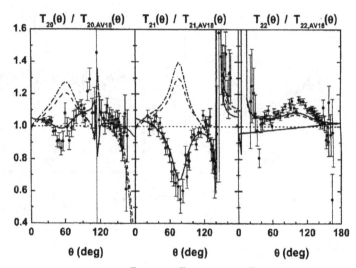

Fig. 10.8   The calculated $\frac{T_{20}}{T_{20,AV18}}$, $\frac{T_{21}}{T_{21,AV18}}$ and $\frac{T_{22}}{T_{22,AV18}}$ with the $Br3NF$ and those with Eq. (10.24), for the $p+d$ scattering at $E_p = 3$ MeV, are compared with experimental data. The dash-dotted lines and the solid lines are for interactions ($V_T = 20$ MeV, $V_0 = -38$ MeV) and ($V_T = -20$ MeV, $V_0 = -45$ MeV). The dashed lines are the same as those in Fig. 10.7.

Next, we will change the sign of V_T so that $V_T = -20$ MeV, and V_0 is readjusted as $V_0 = -45$ MeV to reproduce the ^{3}He binding energy. The interaction with these new parameters gives good agreement between the calculated and the measured values for T_{20} and T_{21}, as seen in the figure. The calculated value of T_{22} is almost unchanged. This success provides a foundation for an improvement of the sign of the tensor interaction.

10.5   Summary and Future Problem in Few Nucleon Systems

The two-nucleon force $AV18$ shows that the binding energies of ^{3}H and ^{3}He are smaller than the empiricals by an order of 1 MeV, and yields a larger doublet cross section $\sigma_1(\theta)$ of the $p + d$ scattering, when compared with that measured in the middle angular region at $E_p = 3$ MeV. Such discrepancies between the $2NF$ calculation and experimental data are solved by taking into account the three-nucleon force, $Br3NF$, which provides a reasonable correction to the binding energies, and improves the cross section satisfactorily. However, difficulties are encountered when the calculations are used to explain the analyzing-power data of the $p + d$ scattering. The values of the vector analyzing powers of the proton, A_y and of the deuteron, iT_{11}, as calculated by the $AV18$ interaction, are much smaller than those obtained by measurement, and the $Br3NF$ interaction hardly improves the calculation. The phenomenological spin-orbit three-nucleon force does fit the data of the vector analyzing powers, but its success is limited to one particular scattering amplitude, S_3. Further, the physical background of such a spin-orbit three-nucleon force is not clear at present. The introduction of the $Br3NF$ into the calculation of the tensor analyzing powers T_{20} and T_{21}, of the $p + d$ scattering, increases the difference between the theoretical and the measured values, and the poor fit of the data suggests the need to change the sign of the tensor part of the employed three-nucleon force.

In recent studies of four-nucleon systems, difficulties similar to those observed in the $p+d$ scattering have been found. In the $p+{}^3$He scattering, the vector analyzing power of the proton $A_y(p)$ calculated by the $AV18$ two-nucleon force is much smaller than that measured at $E_p = 4.05$ MeV as shown in Fig. 10.9 [51]. The calculation by the two-nucleon force, which is described by the dashed line, is about 50% of the measured value at $\theta = 90°$. This discrepancy is not solved by including the three-nucleon force of the Urbana model, as is shown by the solid line. The numerical calculations were carried out by the variational method with the hyper-spherical harmonics (HH) technique, but the result is quite independent of the computational technique, since the same results have been obtained by other methods [51]. In this scattering, a similar discrepancy has been found for the vector analyzing power of ^{3}He, A_y (^{3}He). In these circumstances, the discrepancy between calculated values and experimental data in the vector analyzing power is now one of the most important subjects in the study of the few nucleon systems and is called the "A_y puzzle".

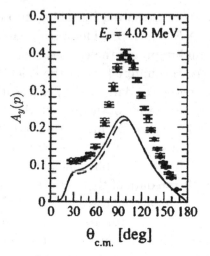

Fig. 10.9   The vector analyzing power of the proton $A_y(p)$ at $E_p = 4.05$ MeV. The dashed line is the calculation by the $AV18$ and the solid line includes the Urbana model $3NF$. The calculations are performed by the HH variation method.

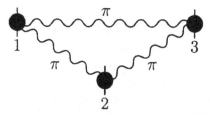

Fig. 10.10   Nuclear force by 3π exchange mechanism. The closed circles express the $I = T = \frac{3}{2}$ resonance.

To solve these difficulties, one will need to investigate the contributions of interactions which have not been examined so far. One potential candidate of such interactions is the three-nucleon force induced by exchanges of three pions between three nucleons, which is illustrated in Fig. 10.10 [52]. In that figure, the pion emitted by nucleon 1 is scattered by nucleon 2, then scattered by nucleon 3, and finally absorbed again by nucleon 1. Since the emission and absorption of the pion by nucleon 1 will be viewed as the scattering of the pion by nucleon 1, the above mechanism is equivalent to

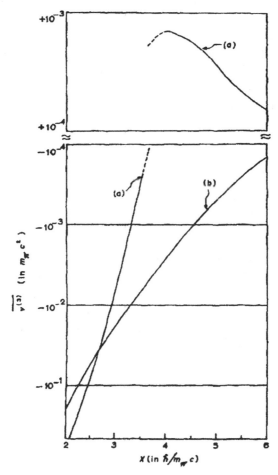

Fig. 10.11 Three-nucleon potential due to the 3π exchange mechanism (a) and the 2π exchange one (b) for a triton in the equilateral triangle configuration. $x = \frac{1}{3}(r_{12} + r_{23} + r_{31})$, where r_{ij}, is the distance between the nucleons i and j.

scattering of the pion three times. These successive scatterings of the pion can be treated by the dispersion relation, and the contribution of each scattering to the interaction is enhanced by the $\left(\frac{3}{2}, \frac{3}{2}\right)$ resonance. This is due to the excitation of the related nucleon, which is similar to the case of the two-pion exchange mechanism. The three-pion (3π) exchange force derived in such a way might be weak due to a higher-order correction when compared with that induced by the two-pion

(2π) exchange mechanism, from the viewpoint of the perturbation theory. However, the contribution of the 3π exchange mechanism will be enhanced by resonance three times in one diagram, and will give finite corrections.

The contribution to the interaction due to the 3π exchange mechanism has been calculated and compared with that of the 2π exchange mechanism, for a triton in the equilateral triangle configuration [52]. The numerical results are shown in Fig. 10.11, where the 3π (a) and the 2π (b) exchange potentials are shown as functions of $x = \frac{1}{3}(r_{12} + r_{23} + r_{31})$, where r_{ij} is the distance between the nucleons i and j. In the region $x \lesssim 2.5\ \hbar/m_\pi c$, the 3π exchange potential is more attractive than that of the 2π, and for $x \gtrsim 3.75\ \hbar/m_\pi c$, the 3π exchange potential is repulsive. This means that the 3π exchange potential has a tendency to increase the 2π exchange attractive force at small x but to decrease when x is larger. Such features of the 3π exchange force may bring about appreciable effects on the $3NF$ in realistic models. If that is so, it would be worthwhile to carry out more detailed analyses on the 3π exchange $3NF$, which would include the spin dependence of interactions.

Appendix

Scattering Amplitudes for T_R- and T_L-Tensor Interactions

In Sec. 4.3, we have presented the scattering amplitudes T_α and T_β for the tensor interaction of the $S = 1$ projectile,

$$T_\alpha = B + D = \frac{1}{\sqrt{2}} \sin\theta \{\sqrt{6}\cos\theta\ F(1120, \cos\theta) + F(1121, \cos\theta)\}$$

(A.1)

and

$$T_\beta = C + \frac{1}{2\sqrt{2}}(B + D)\cot\frac{\theta}{2}$$

$$= \frac{1}{2}\text{co }\theta^2\frac{\theta}{2}\{\sqrt{6}F(1120, \cos\theta) + F(1121, \cos\theta)\}. \quad \text{(A.2)}$$

In this Appendix, we will show that they have the following nature:

$$\text{for } T_L\text{-tensor interaction,} \quad T_\alpha = 0 \quad \text{(A.3)}$$

and

$$\text{for } T_R\text{-tensor interaction,} \quad T_\beta = 0. \quad \text{(A.4)}$$

These suggest that, when the tensor interaction consists of only two types, T_R and T_L, the amplitude T_α represents the scattering by the T_R interaction, and T_β represents the scattering by the T_L interaction. In the following, we will derive Eqs. (A.3) and (A.4) in the plane wave Born approximation (PWBA).

Using the explicit form of T_R where

$$T_R = \sum_\kappa (-)^\kappa S_{2-\kappa} R_{2\kappa} V_R(R) \quad \text{with} \quad S_{2-\kappa} = [s \times s]^2_{-\kappa}$$

$$\text{and} \quad R_{2\kappa} = R^2 Y_{2\kappa}(\hat{R}), \quad (A.5)$$

one can calculate the transition matrix element between the $|s\nu_i\rangle$ and $|s\nu_f\rangle$ states in the $PWBA$ as

$$\langle s\nu_f, \mathbf{k}_f | T_R | s\nu_i, \mathbf{k}_i \rangle = (-)^{s-\nu_f}(ss\nu_i - \nu_f \mid 2\kappa) Y_{2\kappa}(\hat{q}) A(q), \quad (A.6)$$

where

$$q = \mathbf{k}_i - \mathbf{k}_f \quad (A.7)$$

$$A(q) = -\frac{4\pi}{\sqrt{5}} \frac{1}{\sqrt{2s+1}} (s\|[s \otimes s]^2\|s) \int j_2(qR) V_R(R) R^4 dR.$$

$$(A.8)$$

On the other hand, the matrix element is given by the invariant amplitude method so that

$$\langle s\nu_f, \mathbf{k}_f | T_R | s\nu_i, \mathbf{k}_i \rangle$$

$$= (-)^{s-\nu_f}(ss\nu_i - \nu_f \mid 2\kappa) \sum_{l_i=0} [\dot{C}_{l_i}(\hat{k}_i) \otimes C_{l_f=2-l_i}(\hat{k}_f)]^2_x$$

$$\times F(ss2l_i, \cos\theta). \quad (A.9)$$

Comparison of (A.6) with (A.9) gives

$$Y_{2\kappa}(\hat{q}) A(q) = \sum_{l_i=0}^{2} [\dot{C}_{l_i}(\hat{k}_i) \otimes C_{2-l_i}(\hat{k}_f)]^2_\kappa F(ss2l_i, \cos\theta). \quad (A.10)$$

Choosing the coordinate axes, $z \parallel \mathbf{k}_i$ and $y \parallel \mathbf{k}_i \times \mathbf{k}_f$, we get, for $\kappa = 1$ and 2

$$\sqrt{\frac{15}{8\pi}} \cos\theta_q \sin\theta_q \, A(q) = -\sqrt{\frac{3}{2}} \cos\theta \sin\theta \, F(1120, \cos\theta)$$

$$-\frac{1}{2}\sin\theta \, F(1121, \cos\theta) \quad (A.11)$$

and

$$\sqrt{\frac{15}{32\pi}} \sin^2 \theta_q \, A(q) = \sqrt{\frac{3}{8}} \sin^2 \theta \, F(1120, \cos \theta), \qquad (A.12)$$

where Eq. (A.7) provides

$$\sin^2 \theta_q = \frac{k^2}{q^2} \sin^2 \theta, \qquad (A.13)$$

$$\cos \theta_q \sin \theta_q = \frac{k^2}{q^2}(1 - \cos \theta) \sin \theta \qquad (A.14)$$

with

$$k_i = k_f = k. \qquad (A.15)$$

Solving Eqs. (A.11) and (A.12) by the use of Eqs. (A.13) and (A.14), we get

$$F(1121, \cos \theta) = -\sqrt{6}F(1120, \cos \theta), \qquad (A.16)$$

which gives, by Eq. (A.2),

$$T_\beta = 0. \qquad (A.17)$$

This means that the T_R-tensor interaction does not contribute to T_β. Next, we will show that the amplitude T_α vanishes for the T_L-tensor interaction. Since $\boldsymbol{L}$ is the vector, we write $\boldsymbol{L} = L\boldsymbol{1}$ using the unit vector $\boldsymbol{1}$ of the $\boldsymbol{L}$ direction. Then

$$R_{2q}(\boldsymbol{L}, \boldsymbol{L}) = \sum_m (11mm' \mid 2q)L^2. \qquad (A.18)$$

In the classical concept, $\boldsymbol{L}$ is perpendicular to $\boldsymbol{k}_i$ and $L_z = 0$ due to $z \parallel \boldsymbol{k}_i$. This leads to

$$q \neq \pm 1. \qquad (A.19)$$

The matrix element of the T_L interaction in the spin space is given by

$$\langle s\nu_f | \sum_q (-)^q S_{2-q} R_{2q} V_L(R) | s\nu_i \rangle$$

$$= \sum_q (-)^q \langle s\nu_f | S_{2-q} | s\nu_i \rangle R_{2q} V_L(R),$$

where

$$\nu_f - \nu_i = -q.$$

Combined with Eq. (A.19), this gives $\nu_f - \nu_i \neq \pm 1$. However, the amplitudes B and D are the matrix elements for $\nu_f - \nu_i = \pm 1$. Since the amplitude T_α consists of B and D as Eq. (A.1), the T_L-tensor interaction does not contribute to T_α. This suggests that the amplitude T_α stands for the scattering in the T_R interaction, while the amplitude T_β stands for the scattering in the T_L interaction.

References

1. M. G. Mayer and J. HD. Jensen, Elementary Theory of Nuclear Shell Structure (1955) New York. John Wiley & Sons, Inc., London, Chapman & Hall, Ltd.
2. J. M. Daniels, Oriented Nuclei-Polarized Targets and Beams (1965) Academic Press, New York and London.
3. J. P. Elliott and A. M. Lane, The Nuclear Shell Model (Appendix), Handbuch der Physik XXX1X (1957) Springer-Verlag; M. A. Preston, Physics of the Nucleus (1965) Addison-Wesley Publishing Co.
4. Proc. of the 3rd Intern. Symp. on Polarization Phenomena in Nuclear Reactions, ed. by H. H. Barschall and W. Haeberli, The University of Wisconsin Press (1970).
5. G. G. Ohlsen, Rep. Prog. Phys. 35 (1972) 717.
6. L. I. Schiff, Quantum Mechanics (1955).
7. P. A. M. Dirac, Quantum Mechanics (1943).
8. M. L. Goldberger and K. M. Watson, Collision Theory, John Wiley & Sons, Inc. (1964).
9. F. D. Santos, Nucl. Phys. A236 (1974) 90.
10. M. Tanifuji and K. Yazaki, Prog. Theor. Phys. 40 (1968) 1023.
11. D. J. Hooton and R. C. Johnson, Nucl. Phys. A175 (1971) 583.
12. B. Robson, The Theory of Polarization Phenomena, Clarendon Press, Oxford (1974).
13. M. Matsuoka et al., Nucl. Phys. A455 (1986) 418.
14. G. Tungate et al., Phys. Lett. 98B (1981) 347.
15. P. E, Hodgson, The Optical Model of Elastic Scattering, Clarendon Press, Oxford (1963).
16. F. D. Becchetti and G. W. Greenlees, Phys. Rev. 182 (1969) 1190.
17. J. J. H. Menet, E. E. Gross, J. J. Malanify and A. Zucker, Phys. Rev. C4 (1971) 1114.
18. L. Wolfenstein, Annual Review of Nuclear Science, 6 (1956) 43.
19. H. Sakaguchi et al., Phys. Rev. C26 (1982) 944, J. Phys. Soc. Jpn. 55 Suppl. (1986) 99.

20. M. Yahiro, Y. Iseri, H. Kameyama, M. Kamimura and M. Kawai, Prog. Theor. Phys. Suppl. 89 (1986) 32.
21. R. V. Reid, Ann. of Phys. 50 (1968) 411.
22. Y. Iseri, H. Kameyama, M. Kamimura, M. Yahiro and M. Tanifuji, Nucl. Phys. A490 (1988) 375.
23. Y. Iseri and M. Tanifuji, Phys. Lett. B354 (1995) 183.
24. H. Hamma *et al.*, Phys. Rev. C41 (1990) 2737.
25. Y. Iseri, M. Tanifuji, H. Kameyama, M. Mamimura and M. Yahiro, Nucl. Phys. A533 (1991) 574.
26. H. Ohnishi, M. Tanifuji, M. Kamimura, Y. Sakuragi and M. Yahiro, Nucl. Phys. A415 (1984) 271.
27. Z. Moroz *et al.*, Nucl. Phys. A381 (1982) 294.
28. Y. Sakuragi, M. Yahiro and M. Kamimura, Prog. Theor. Phys. Suppl. 89 (1986) 136.
29. G. R. Satchler and W. G. Love, Phys. Reports 55 (1979) 183.
30. G. Tungate *et al.*, J. Phys. G12 (1986) 1001.
31. Y. Hirabayashi, Y. Sakuragi and M. Tanifuji, Phys. Lett. B318 (1993) 32.
32. Y. Sakuragi, M. Yahiro, M. Kamimura and M. Tanifuji, Nucl. Phys. A462 (1987) 173.
33. M. Tanifuji and H. Kameyama, Phys. Rev. C60 (1999) 034607.
34. M. Yamaguchi *et al.*, Phys. Rev. C74 (2006) 064606.
35. M. Tanifuji, M. Yamaguchi and H. Kameyama, Phys. Rev. C79 (2009) 027604.
36. M. Glor *et al.*, Nucl. Phys. A286 (1977) 31.
37. M. Geso *et al.*, Phys. Rev. C65 (2002) 034005.
38. R. H. Landau and M. Sagen, Phys. Rev. C33 (1986) 447.
39. M. A. Franey and W. G. Love, Phys. Rev. C31 (1985) 488.
40. H. T. Coelho, T. K. Das and M. R. Robilotta, Phys. Rev. C28 (1983) 1812.
41. S. Ishikawa, M. Tanifuji, Y. Iseri and Y. Yamamoto, Phys. Rev. C72 (2005) 027601.
42. J. Fujita and H. Miyazawa, Prog. Theor. Phys. 17 (1957) 366.
43. L. D. Faddeev, Soviet Phys. JETP12 (1961) 1014.
44. S. Ishikawa, Few-Body Syst. 32 (2003) 229. References therein.
45. R. B. Wiringa, V. G. J. Stoks and R. Schiavilla, Phys. Rev. C51 (1995) 238.
46. S. Ishikawa, M. Tanifuji and Y. Iseri, Phys. Rev. C66 (2002) 044005.
47. K. Sagara *et al.*, Phys. Rev. C50 (1994) 576; S. Shimizu *et al.*, Phys. Rev. C52 (1995) 1193.
48. A. Kievsky, Phys. Rev. C60 (1999) 034001.
49. S. Ishikawa, M. Tanifuji and Y. Iseri, Phys. Rev, C67 (2003) 061001 (R).
50. S. Ishikawa, M. Tanifuji and Y. Iseri, Few-Body Syst, 17 (2004) 561.
51. B. M. Fisher *et al.*, Phys. Rev. C74 (2006) 034001.
52. J. Fujita, M. Kawai and M. Tanifuji, Nucl. Phys. 29 (1962) 252.